Copyright

ISBN 978-1-7389501-8-8

Acknowledgements

"At the age of eleven, I began Euclid, with my brother as tutor. This was one of the great events of my life, as dazzling as first love. I had not imagined there was anything so delicious in the world. From that moment until I was thirty-eight, mathematics was my chief interest and my chief source of happiness."

– *Bertrand Russell, Autobiography*

"Mathematicians, like Proust and everyone else, are at their best when writing about their first love."

– *Gian-Carlo Rota, Discrete Thoughts*

I express my gratitude to:

- The Almighty Creator, for providing me with this blessed and privileged life.
- My parents, for financing my mathematical education, and for supporting me during the time that this book was written.
- My students, for asking wonderful questions.
- My friends, for their companionship and for listening to me talk about mathematics.
- Euclid, for writing the *Elements*, which showed the world the meaning of eternal rigour.

Special thanks are extended to Warren Bei for reading the the first edition's manuscript and offering numerous suggestions, many of which were implemented. Any remaining mathematical errors or mistakes in the typesetting are my responsibility alone.

No artificial intelligence was used in the writing of the material for this book. Occasional assistance was sought from large language models for surmounting typesetting and formatting challenges in LaTeX.

Contents

Preface

> "In studying a philosopher, the right attitude is neither reverence nor contempt, but first a kind of hypothetical sympathy, until it is possible to know what it feels like to believe in his theories, and only then a revival of the critical attitude, which should resemble, as far as possible, the state of mind of a person abandoning opinions which he has hitherto held. Contempt interferes with the first part of this process, and reverence with the second. Two things are to be remembered: that a man whose opinions and theories are worth studying may be presumed to have had some intelligence, but that no man is likely to have arrived at complete and final truth on any subject whatever."
>
> – *Bertrand Russell, A History of Western Philosophy*

Mathematics is the study of ultimate regularity. Regularity entails order or predictability. Its antithesis is chaos. When there is regularity, there are discernible objects at play. In other words, there is structure. Wherever there is structure, there is symmetry. Symmetry means that, while one aspect of the object changes, another remains unchanged. The present book is an effort to rigorously systematize and provide an exposition of those aspects of elementary algebra that appear in school curricula and mathematical contests. In the course of writing, it became evident that there are three recurring themes among the proof techniques used, all of which are forms of symmetry:

A. The discrete Fubini's principle instructs us to write the same object in two different ways. There are many examples of how we have applied this principle:

   1. There is a theorem dedicated to the ways in which one may switch from iterating through the rows of a matrix to columns of the matrix or vice versa (Theorem 4.18).
   2. It is used to prove the Girard-Newton sums (Theorem 11.53).
   3. It is used in the proof of Chebyshev's inequality (Corollary 12.19).
   4. The presented proof of Hölder's inequality (Theorem 12.25) utilizes it.

B. Antisymmetry in a partial order is a powerful method of proof that lets us break down the strong notion of equality into the conjunction of two individually weaker statements. Examples of antisymmetry that we have used in this book are:

   1. Set equality can be broken into two subset relations, specifically $A = B$ if and only if both $A \subseteq B$ and $B \subseteq A$ (Theorem 2.3).
   2. Real number equality can be split into two inequality relations, meaning $x = y$ if and only if both $x \leq y$ and $y \leq x$ (Definition 5.39).

C. Applying an equivalence relation to a set groups the distinct elements into equivalence classes, where elements in a class are equated. This allows us to focus on the essential properties of objects which are preserved under the relation, and to choose a convenient canonical representative of each class.

   1. Propositional variables are logically equivalent if they have identical final columns in a joint truth table (Definition 1.18).
   2. Real or complex numbers are equal to each other if they are exactly the same number.
   3. Functions are equal to each other if they have the same domain, and map the same inputs to the same outputs (Definition 2.24).
   4. Formal polynomials are equal if they have the same sequence of coefficients (Definition 11.4).

It is our hope that the reader will keep these proof techniques in mind while reading the book, and that the impression of the importance of symmetry will grow as the reader encounters the methods time and again.

The intended audience consists of students of math contests, competitions, and olympiads who want to take a rigorous second look at the results that they might be accustomed to taking for granted, and teachers, coaches, and trainers who want to reinforce their own understanding of what they teach.

Suggestions, comments, and error submissions would be greatly appreciated. These could include ideas for strengthening or generalizing theorems and other additional material, and fixing mathematical or typographical mistakes. Messages may be sent to

`academy@existsforall.com`

Samer Seraj
Mississauga, Ontario, Canada
June 6, 2026

Chapter 1

Logic

"Faith is different from proof; the latter is human, the former is a Gift from God."

– *Blaise Pascal*

"The more I think about language, the more it amazes me that people ever understand each other at all."

– *Kurt Gödel*

"A person, who had not mastered the art of the proofs in high school, is as a rule unable to distinguish correct reasoning from that which is misleading. Such people can be easily manipulated by the irresponsible politicians."

– *Vladimir Arnol'd*

In order to explore mathematical topics, we need to have an outline of how mathematical exploration works. The reader should already have an intuitive sense of mathematics, so now we will focus on that which allows us to verify our intuition: logic. Formal logic is more complicated than we will let on. We will approach the topic with the practical aim of familiarizing the reader with what is necessary to engage in actual mathematical practice, as opposed to formally analyzing mathematical practice itself. What the reader should learn from our exposition is how to jump from hypotheses to conclusions using rules of inference, and how to show that two propositions are logically equivalent.

1.1 Propositions

Definition 1.1. A **definition** is a clear description of an object or concept (in a circular fashion, this is the definition of a definition!).

Unfortunately, essentially undefined terms are inevitable, such as when we start to work the fundamental objects of the mathematical universe, called sets, in Definition 2.1.

Definition 1.2. A **proposition** makes a declaration that we can understand and is not nonsense. There are two "laws of thought" about propositions that are relevant to logic:

- Law of excluded middle: Each proposition is true or its negation is true.
- Law of non-contradiction: There is no proposition such that both it and its negation is true.

Thus, each proposition has a *unique* **truth value** of true or false. We will justify these two laws of thought using truth tables in Theorem 1.6. To be clear, a proposition is a statement that does not have any variables and instead deals with fixed objects, like numbers; we will define statements that have variables in Definition 1.35. A **propositional variable** is a variable that represents propositions (which is different from our usual variables that represent objects like numbers). Propositions are usually denoted by capital letters like P, Q, R and propositional variables are usually denoted by small letters like p, q, r.

Example. The proposition $1+1=2$ is a proposition that is true. The proposition $2+2=5$ is a proposition that is false. Notice that we do not have variables representing numbers in these propositions, like in $x > 0$.

We will not delve too precisely into what exactly is allowed or not allowed in the construction of a proposition. Rather, one should become accustomed to constructing and interpreting propositions by considering their meanings, just like when we write or read in natural language without consciously worrying about grammar.

As we come across sets, numbers, geometric spaces, and other mathematical objects in the text, we show the reader by example what propositions are acceptable about each. However, a cursory overview of the underlying language of logic will be necessary for us to proceed.

Definition 1.3. Propositions can be used to produce other propositions via any or all of the following three logical operators, whose meanings correspond to our intuition:

- **Negation**: "not" is a unary operator, meaning it operates on a single proposition P, and its application is denoted by $\neg P$.
- **Conjunction**: "and" is a binary operator, meaning it operates on a pair of propositions P and Q, and its application is denoted by $P \wedge Q$.
- **Disjunction**: "or" is another binary operator, and its application is denoted by $P \vee Q$.

The propositional variables $\neg p, p \wedge q, p \vee q$ have the following **truth tables** which will be useful in defining proofs later. We emphasize that truth tables are not proofs themselves. Also, truth tables are of propositional variables, not propositions themselves. This is just like a variable in algebra is not a number itself.

p	q	$p \wedge q$	$p \vee q$
0	0	0	0
0	1	0	1
1	0	0	1
1	1	1	1

p	$\neg p$
0	1
1	0

The reader should go through the truth tables and check that each line matches one's expectations based on our intuitive understanding of the three operators in natural language. For example, it makes sense that P is true if and only if $\neg P$ is false, and that P is false if and only if $\neg P$ is true. Combining propositions using these operators produces a **compound proposition**, where the constituent propositions are called **components**.

Problem 1.4. For each proposition P, let $t(P)$ denote 0 or 1 according to whether P is false or true, respectively. It may be verified that

$$t(p \wedge q) = \min(t(p), t(q)),$$
$$t(p \vee q) = \max(t(p), t(q)).$$

Find a way of expressing $t(p \wedge q)$ and $t(p \vee q)$ by applying some ordinary arithmetic operations (addition, subtraction, multiplication, and division) to $t(p)$ and $t(q)$. Do the same for $t(\neg p)$ in terms of $t(p)$.

Definition 1.5. A propositional variable that has a value of 1 or "true" for all truth assignments of 0's and 1's to its components is called a **tautology**. On the other hand, a proposition that has a value of 0 or "false" for all truth assignments of 0's and 1's to its components is called a **contradiction**. A proposition that has neither all 0's nor all 1's in the rightmost column of its truth table is called a **contingency**, though this term is not commonly used. We will denote a tautology by $\top$ and a contradiction by $\bot$.

Theorem 1.6. The following assertions hold ("hold" means to be true):

1. The disjunction $p \vee \neg p$ is a tautology. This justifies the law of excluded middle.
2. The conjunction $p \wedge \neg p$ is a contradiction. This justifies the law of non-contradiction.

Proof. We will show that these results hold by constructing truth tables using the definitions of logical operators.

p	$\neg p$	$p \vee \neg p$	$p \wedge \neg p$
0	1	1	0
1	0	1	0

Thus, $p \vee \neg p$ is always true and $p \wedge \neg p$ is always false. □

Definition 1.7. There are three further common logical operators that can be pieced together using $\neg, \wedge, \vee$:

- **Exclusive or** (XOR): $P \oplus Q$ denotes either of the propositions

$$(P \vee Q) \wedge \neg(P \wedge Q),\ (P \wedge \neg Q) \vee (Q \wedge \neg P)$$

  that lead to equivalent propositional variables (we will define equivalence in Definition 1.18)

- **Implication**: $P \to Q$ denotes $\neg P \vee Q$
- **Biconditional**: $P \leftrightarrow Q$ denotes $(P \to Q) \wedge (Q \to P)$

These operators lead to the following truth tables for the corresponding propositional variables:

p	q	$p \oplus q$	$p \to q$	$p \leftrightarrow q$
0	0	0	1	1
0	1	1	1	0
1	0	1	0	0
1	1	0	1	1

Those familiar with modular arithmetic (see Volume 3) will recognize that the $\oplus$ operator is the same as addition modulo 2. Note that $p \oplus q$ and $\neg(p \leftrightarrow q)$ have the same truth table. Also, unless otherwise specified, the word "or" refers to the ordinary (inclusive) $\vee$ and not the exclusive $\oplus$.

Definition 1.8. Just like with arithmetic operations, we need to establish an order in which logical operators are interpreted in a proposition or propositional variable, unless we wish to be swarmed by parentheses. To this end, we rank them from highest to lowest, though it never hurts to add some clarifying parentheses that are technically unnecessary: $\neg$, then $\vee, \wedge$ are of equal rank, then $\to, \leftrightarrow$ are also of equal rank. Any operators of equal rank will need parentheses in order for us to know the order of interpretation. This raises the question of whether we can avoid parentheses in some of those cases as well, like with arithmetic operations. We will answer this question momentarily. The symbol $\oplus$ is omitted from this list because it is rarely used and we instead say something in words like "P or Q but not both."

Our ultimate goal here is to be able to claim that certain propositions are true, under the assumption that other propositions are true. We will have to make it clear what this means in order to pursue it.

Definition 1.9. There are propositions that are assumed to be true, which are called **axioms**. We discuss some of the axioms for the mathematical universe of sets in Chapter 2. Axioms are necessary because logic works in a way where each truth is assumed or is derived from other truths, and no truth is absolute. A proposition whose truth value is unknown is called a **conjecture**.

Definition 1.10. An **argument** or **derivation** is a finite sequence of propositions. One may formally write an argument with commas like $P_1, P_2, \ldots, P_n, Q$ or stack them from top to bottom. The final proposition Q is called the **conclusion** or **consequence**, and all other propositions P_i in the sequence are called **hypotheses** or **premises**.

- Intuitively, an argument is said to be **valid** if the assumption that the premises are true forces the conclusion to be true. The question remains as to what it mean to "force" the conclusion to be true from the truth of the hypotheses. Here is the absolutely critical idea: complete a joint truth table that has a column for each of the propositions $P_1, P_2, \ldots, P_n, Q$, and substitute in all possible combinations of 0's and 1's for the components of P_i and Q, just like in truth tables for propositional variables. If the value for the conclusion Q is 1 in every row where all of the premises P_i have value 1, then we have a valid argument.
- A **proof** or **sound argument** is a valid argument where we may assume that the premises are true, because each premise is either an axiom (and therefore assumed to be true) or a conclusion of another valid argument where all of the premises are true. We say that axioms and conclusions of sound

arguments are **true**, and that their negations are **false**. A conclusion whose truth follows from a proof in this way is called a **result**, **theorem**, **principle**, or **postulate**, to list a few names. The word "theorem" is usually reserved for particularly important results. A result that is used to prove a theorem is called a **lemma**, and a result that a theorem is used to prove is called a **corollary**.

So, checking the validity of an argument is like completing the truth table for the shells or skeletons of its constituent propositions, where it is not the particular premises and conclusions that matter, but their *form*. Thus, for a proposition S, we call s its **form** if s looks exactly like S symbolically but s is a propositional variable. For example, the form of $X \vee Y$ can be written as $x \vee y$.

Example 1.11. Determine whether the following argument is valid: $P \vee (\neg Q \to R), \neg R, P \vee Q$.

Solution. We construct a truth table as follows:

p	q	r	$\neg q$	$\neg q \to r$	$p \vee (\neg q \to r)$	$\neg r$	$p \vee q$
0	0	0	1	0	0	1	0
0	0	1	1	1	1	0	0
0	1	0	0	1	$\boxed{1}$	$\boxed{1}$	$\boxed{1}$
0	1	1	0	1	1	0	1
1	0	0	1	0	$\boxed{1}$	$\boxed{1}$	$\boxed{1}$
1	0	1	1	1	1	0	1
1	1	0	0	1	$\boxed{1}$	$\boxed{1}$	$\boxed{1}$
1	1	1	0	1	1	0	1

In every row where $p \vee (\neg q \to r)$ and $\neg r$ are true, $p \vee q$ is also true. Thus, the argument is valid. □

Definition 1.12. In a valid argument P, Q, we say that Q is a **necessary condition** for P, and that P is a **sufficient condition** for Q. If P, Q is a valid argument, then we say anything of the following: "P implies Q," or "Q is implied by P," or "if P then Q," or "Q if P," or "P only if Q." Unfortunately, it is commonplace in definitions to say "if" when we really mean "if and only if" like a biconditional; so, definitions are biconditional unless otherwise specified.

Definition 1.13. There are three propositions related to an implication $P \to Q$ that we will frequently need:

- **Converse**: $Q \to P$
- **Contrapositive**: $\neg Q \to \neg P$
- **Inverse**: $\neg P \to \neg Q$

If $P \to Q$ is true, it does not mean that all three of the others are true. The contrapositive is guaranteed to be valid because the the form of the contrapositive is equivalent to the origin implication, as we will show in Theorem 1.28, but not the other two. However, if either the converse or inverse is true, then both are true because the converse and inverse and contrapositives of each other. The term "inverse" is not commonly used in our experience, so we will usually use a more descriptive name, like "the contrapositive of the converse." This also prevents confusion with the myriad of other inverses in math.

Example. Let P be the proposition "I am holding an apple" and Q be the proposition "I am holding a fruit." Then $P \to Q$ is true because if I am holding an apple, then I am definitely holding a fruit. However, we cannot claim that $Q \to P$ is true because if I am holding a fruit, then that fruit might not be an apple. Notice that the contrapositive $\neg Q \to \neg P$ is true though, as it correctly states that if I am not holding a fruit, then I am not holding an apple.

Problem 1.14. Prove that the following arguments are not valid:

1. Converse fallacy: $P \to Q, Q, P$
2. Inverse fallacy: $P \to Q, \neg P, \neg Q$
3. Non-equivalence: $P \to Q, Q \to P$
4. Begging the question: $Q \to Q, Q$
5. Circular reasoning: $Q \to P, P \to Q, Q$

These are all common mistakes, so take heed.

Definition 1.15. It is clear that filling out truth tables in this manner is an exhausting activity. We would benefit from developing a collection of valid arguments that are widely applicable and preferably intuitively understandable too (and therefore possible to remember with ease). We will call these **rules of inference** and they are the subjects of Theorem 1.16 and Theorem 1.26.

Theorem 1.16. The following are the most common rules of inference, excluding what we can derive from logical equivalences using Theorem 1.26:

Name	Rule of Inference
Addition	$P, P \vee Q$
Simplification	$P \wedge Q, P$
Conjunction	$P, Q, P \wedge Q$
Modus Ponens	$P, P \to Q, Q$
Modus Tollens	$\neg Q, P \to Q, \neg P$
Hypothetical Syllogism (implication transitivity)	$P \to Q, Q \to R, P \to R$
Disjunctive Syllogism (elimination)	$\neg P, P \vee Q, Q$
Constructive Dilemma	$P \to R, Q \to S, P \vee Q, R \vee S$
Destructive Dilemma	$P \to R, Q \to S, \neg R \vee \neg S, \neg P \vee \neg Q$
Biconditional Elimination	$P \leftrightarrow Q, P \vee Q, P \wedge Q$ $P \leftrightarrow Q, \neg P \vee \neg Q, \neg P \wedge \neg Q$
Bidirectional Dilemma	$P \to R, Q \to S, P \vee \neg S, R \vee \neg Q$

The constructive dilemma begets several children. One special case is resolution, where Q is $\neg P$, in which case $P \vee Q$ is a tautology that can be removed. This yields the rule

$$P \to R, \neg P \to S, R \vee S.$$

Another special case is disjunctive elimination, which has R and S being the same, which yields

$$P \to R, Q \to R, P \vee Q, R.$$

At the intersection of the last two, we get

$$P \to Q, \neg P \to Q, Q.$$

There is no need to memorize these names, as it is far more important to become accustomed to using these rules. Some of these rules generalize to longer arguments. Instead of listing them here, we allow them to come up in practice. The reader should independently think about such generalizations as well.

Corollary 1.17 (Principle of explosion). If both P and $\neg P$ are true (in some hypothetical scenario), then every proposition Q is true, which is a highly undesirable scenario. This justifies the law of non-contradiction, as it says that a proposition and its negation cannot both be true.

Proof. Suppose P and true and $\neg P$ is also true. Let Q be any proposition. Then $P \vee Q$ is true by addition. Since $\neg P$ holds, we can apply disjunctive syllogism to get that Q is true. So, in a bizarre world where P can be true and false at the same time, every proposition Q is true. Note that this means $\neg Q$ is also true (since $\neg Q$ is also a proposition), so in fact every proposition is both true and false. A consequence like this that utterly erodes our standard notions of truth and falsity must be prevented from manifesting. □

1.2 Equivalence

Now we come to the important concept of logical equivalences, which, among other advantages, will allow us to construct more rules of inference, thanks to Theorem 1.26.

Definition 1.18. Two propositional variables are said to be **logically equivalent** if they produce the same truth table. By "same," we mean that we draw a joint truth table, and the final two columns corresponding to the two propositional variables need to match. It is necessary for the truth tables to contain all possible assignments of 0's and 1's to all components. We denote equivalence of two propositional variables by placing the symbol $\equiv$ between them. For example, $\neg(\neg p) \equiv p$.

Theorem 1.19. Logical equivalence satisfies the following properties (which we study more generally via equivalence relations in Definition 2.17):

- Reflexive: $p \equiv p$
- Symmetric: $p \equiv q$ implies $q \equiv p$
- Transitive: $p \equiv q$ and $q \equiv r$ together imply $p \equiv r$

Equivalences are very useful for simplifying complex propositional variables or writing them in alternate illuminating forms. If a part of a propositional variable is replaced by a logically equivalent part, then the new propositional variable is equivalent to the old one. This is one of the great strengths of equivalences.

Proof. The properties of reflexivity, symmetry and transitivity should be clear from the definition of logical equivalence. For the replacement property note that if $p \equiv q$, then for any r,

$$\begin{aligned} \neg p &\equiv \neg q, \\ p \vee r &\equiv q \vee r, \\ p \wedge r &\equiv q \wedge r. \end{aligned}$$

Thus, we can prove the replacement property mentioned at the end by starting with $p_1 \equiv p_2$ and noting that wrapping negations, disjunctions, and conjunctions around it step-by-step produces more equivalences. A formal proof of this would use a method called structural induction. □

Theorem 1.20. If two propositions have equivalent forms and one is true in an axiom system, then the other is true too. This makes logical equivalences very useful as, in order to prove a proposition, it suffices to prove a proposition with an equivalent form.

Proof. Suppose P and Q are propositions with equivalent forms in an axiom system, and that P is true, meaning P is the conclusion of a sound argument. Now we draw the truth table of the argument P, Q. Since p and q are logically equivalent, in each row both have 0 or both have value 1. Thus, whenever p has value 1, q also has value 1, making P, Q a valid argument. Moreover, P is the conclusion of a sound argument. Therefore, Q is the conclusion of the sound argument P, Q, making it true. □

Theorem 1.21. An equivalence that is easy to see is that $\neg(\neg p) \equiv p$. The following table provides more such common equivalences. These are useful in reformulating assertions, which can have a real impact on how convenient it is to prove something. Let p, q, r be propositional variables, and $\top$ denote a tautology and $\bot$ denote a contradiction. Then:

Name of Law	Disjunction	Conjunction
Negation	$p \vee \neg p \equiv \top$	$p \wedge \neg p \equiv \bot$
Annihilation	$p \vee \top \equiv \top$	$p \wedge \bot \equiv \bot$
Identity	$p \vee \bot \equiv p$	$p \wedge \top \equiv p$
Idempotent	$p \vee p \equiv p$	$p \wedge p \equiv p$
Absorption	$p \vee (p \wedge q) \equiv p$	$p \wedge (p \vee q) \equiv p$
Commutative	$p \vee q \equiv q \vee p$	$p \wedge q \equiv q \wedge p$
Associative	$p \vee (q \vee r) \equiv (p \vee q) \vee r$	$p \wedge (q \wedge r) \equiv (p \wedge q) \wedge r$
Distributive	$p \vee (q \wedge r) \equiv (p \vee q) \wedge (p \vee r)$	$p \wedge (q \vee r) \equiv (p \wedge q) \vee (p \wedge r)$
De Morgan's	$\neg(p \vee q) \equiv \neg p \wedge \neg q$	$\neg(p \wedge q) \equiv \neg p \vee \neg q$

The associative laws allow us to write a sequence of disjunctions or a sequence of conjunctions without parentheses, but not if disjunctions and conjunctions are mixed. The commutative laws then allow us to write a sequence of disjunctions in any order, and similarly for conjunctions, but again not if they are mixed. When disjunctions and conjunctions both appear, other equivalences like the absorption laws and distributive laws might be helpful.

Corollary 1.22. By repeated applications of de Morgan's laws, these two laws may be extended to the disjunction or conjunction of many propositions:

$$\neg(p_1 \vee p_1 \vee \cdots \vee p_n) \equiv \neg p_1 \wedge \neg p_2 \wedge \cdots \wedge \neg p_n,$$
$$\neg(p_1 \wedge p_1 \wedge \cdots \wedge p_n) \equiv \neg p_1 \vee \neg p_2 \vee \cdots \vee \neg p_n.$$

Also, note that de Morgan's laws imply that only one of $\vee$ or $\wedge$ needs to be taken as "atomic" in the sense that one of them can be written in terms of the other and negation.

Problem 1.23. Using truth assignments, prove that, unlike disjunction and conjunction, implication is not associative. That is, $p \to (q \to r)$ does not always have the same truth value as $(p \to q) \to r$. Thus, it is technically ambiguous to write $p \to q \to r$ without parentheses. In practice, however, this notation is often used to denote that

$$(p \to q) \wedge (q \to r).$$

In fact, the notation extends to finite sequences of conjunctions of implications, as well as finite sequences of biconditionals; implications and biconditionals can be mixed like $p \to q \leftrightarrow r \to s$. If there is any chance of confusion when using such notation, we will make clarifications.

Lemma 1.24. An argument $P_1, P_2, \ldots, P_n, Q$ is valid if and only if

$$(p_1 \wedge p_2 \wedge \cdots \wedge p_n) \to q$$

is a tautology, where p_i is the form of P_i for each i and q is the form of Q. As a result, the implication operator is of paramount importance in proofs.

Proof. By de Morgan's laws, we first note that the propositional variable

$$(p_1 \wedge p_2 \wedge \cdots \wedge p_n) \to q$$

is a tautology if and only if the equivalent propositional variable

$$\neg p_1 \vee \neg p_2 \vee \cdots \vee \neg p_n \vee q$$

is a tautology, meaning it has a truth table whose rightmost column has all 1's.
For one direction, suppose $P_1, P_2, \ldots, P_n, Q$ is valid. If any of the p_i have a value of 0, then

$$\neg p_1 \vee \neg p_2 \vee \cdots \vee \neg p_n \vee q$$

automatically has a value of 1. If all of the p_i have a value of 1, then the validity of $P_1, P_2, \ldots, P_n, Q$ implies that q also has a value of 1, which is all we need.
For the other direction, suppose

$$\neg p_1 \vee \neg p_2 \vee \cdots \vee \neg p_n \vee q$$

is a tautology, meaning its truth table has a rightmost column that is filled with all 1's. Suppose all of $p_1, p_2, \ldots, p_n$ have values of 1 in some row. Then their negations all have value 0, which forces q to have value 1 in that row, as desired, otherwise $\neg p_1 \vee \neg p_2 \vee \cdots \vee \neg p_n \vee q$ would not have a value of 1 in that row. □

Lemma 1.25. For two propositional variables p and q, it holds that $p \equiv q$ if and only if $p \leftrightarrow q$ is a tautology.

Proof. For one direction, suppose p and q are logically equivalent. Then the truth tables of p and q match. The following table holds:

p	q	$p \to q$	$q \to p$	$p \leftrightarrow q$
0	0	1	1	1
1	1	1	1	1

Every line of the truth table of $p \leftrightarrow q$ is identical to either of these two lines because the truth values of p and q always match due to their logical equivalence. Since the rightmost column is made up of all 1's, $p \leftrightarrow q$ is a tautology.
For the other direction, suppose $p \leftrightarrow q$ is a tautology. We will now use the method of contradiction that we will shortly describe. Suppose, for contradiction, that p and q are not equivalent. This means, somewhere along the line, the value of p is 0 and the value of q is 1, or the value of p is 1 and the value q is 0. Then the truth table of $p \leftrightarrow q$ has at least one of the following two rows:

p	q	$p \to q$	$q \to p$	$p \leftrightarrow q$
0	1	1	0	0
1	0	0	1	0

Since the rightmost column has all 0's, this means $p \leftrightarrow q$ is not a tautology, which is a contradiction since we know that $p \leftrightarrow q$ is a tautology. Technically, the method of contradiction that we will derive in Theorem 1.29 might not apply to arguments *about* propositions themselves like this result, but the method of contradiction is fundamental enough that we will assume that we can use it as a method of proof here. □

Theorem 1.26. For two propositions P and Q, it holds that both P, Q and Q, P are valid arguments if and only if the logical equivalence $p \equiv q$ holds. This is a great way of finding rules of inference.

Proof. By Lemma 1.25, $p \equiv q$ is a logical equivalence if and only if $p \leftrightarrow q$ is a tautology. Recall that $p \leftrightarrow q$ is equivalent to

$$(p \to q) \wedge (q \to p,)$$

so this is also a tautology. As a conjunction is true if and only if each of its components is true, $(p \to q) \wedge (q \to p)$ is a tautology if and only if $p \to q$ and $q \to p$ are tautologies in their own right. By Lemma 1.24, this is true if and only if P, Q and Q, P are valid arguments. □

Definition 1.27. It is common practice to use the symbols $\implies$ and $\iff$ in proofs. In practice, saying that $P \implies Q$ is true means that $P \to Q$ is the conclusion of a sound argument, and saying that $P \iff Q$ is true means that $P \leftrightarrow Q$ is the conclusion of a sound argument.

Theorem 1.28. Due to Lemma 1.24, Lemma 1.25, and Theorem 1.26, much of mathematical proof-writing is about understanding the "flow" of implications and biconditionals. To that end, we offer a number of logical equivalences involving implications and biconditionals between propositional variables:

Name of Law	Equivalence
Definition	$p \to q \equiv \neg p \vee q$
Contrapositive	$p \to q \equiv \neg q \to \neg p$
Disjunction	$p \vee q \equiv \neg p \to q$
Conjunction	$p \wedge q \equiv \neg(p \to \neg q)$
Negation	$\neg(p \to q) \equiv p \wedge \neg q$
Absorption	$p \to (p \wedge q) \equiv p \to q$
Exportation/Importation	$(p \wedge q) \to r \equiv p \to (q \to r)$

Type	In Common	Equivalence
Disjunction	Hypothesis	$(p \to q) \vee (p \to r) \equiv p \to (q \vee r)$
Disjunction	Conclusion	$(p \to r) \vee (q \to r) \equiv (p \wedge q) \to r$
Conjunction	Hypothesis	$(p \to q) \wedge (p \to r) \equiv p \to (q \wedge r)$
Conjunction	Conclusion	$(p \to r) \wedge (q \to r) \equiv (p \vee q) \to r$

Name of Law	Equivalence
Implication and Converse	$p \leftrightarrow q \equiv (p \to q) \wedge (q \to p)$
Contrapositive and Inverse	$p \leftrightarrow q \equiv (\neg q \to \neg p) \wedge (\neg p \to \neg q)$
Implication and Inverse	$p \leftrightarrow q \equiv (p \to q) \wedge (\neg p \to \neg q)$
Contrapositive and Converse	$p \leftrightarrow q \equiv (\neg q \to \neg p) \wedge (q \to p)$
Factored Decomposition	$p \leftrightarrow q \equiv (p \wedge q) \vee (\neg p \wedge \neg q)$
Negation or XOR	$\neg(p \leftrightarrow q) \equiv p \leftrightarrow \neg q \equiv \neg p \leftrightarrow q$

Now we will share the most common methods of proof, which are direct proofs, proof by contrapositive, proof by casework, and several forms of proof by contradiction. In Section 2.3, we introduce several forms of one more method of proof, called induction.

Theorem 1.29 (Methods of proof). A proof that establishes a substantial theorem typically utilize multiple techniques, but we can outline common individual lines of attack as follows.

1. **Direct proof**: The argument
$$H_1, H_2, \ldots, H_n, (P \to Q)$$
is valid if and only if so is the argument
$$H_1, H_2, \ldots, H_n, P, Q.$$

2. **Proof by contrapositive**: The argument
$$H_1, H_2, \ldots, H_n, (P \to Q)$$
is valid if and only if so is the argument
$$H_1, H_2, \ldots, H_n, \neg Q, \neg P.$$

3. **Proof by casework** or **exhaustion**: The argument

$$H_1 \vee H_2 \vee \cdots \vee H_n, P$$

is valid if and only if the n arguments H_i, P for $i = 1, 2, \ldots, n$ are valid.

4. **Proof by contradiction**: The argument

$$H_1, H_2, \ldots, H_n, P$$

is valid if and only if the argument

$$H_1, H_2, \ldots, H_n, \neg P, \bot$$

is valid. Here, $\bot$ refers to a proposition whose form is a logical contradiction, meaning the right-most column of the truth table of the form has all 0's. A special case of this is that the argument $H_1, H_2, \ldots, H_n, P$ is valid if and only if both of the two arguments

$$H_1, H_2, \ldots, H_n, \neg P, Q$$
$$H_1, H_2, \ldots, H_n, \neg P, \neg Q,$$

are valid. A special case of this special case is that $H_1, H_2, \ldots, H_n, P$ is valid if and only if the argument is valid that

$$H_1, H_2, \ldots, H_n, \neg P, P.$$

Note that establishing $P \to Q$ as in direct proofs or proof by contrapositive allows us to then apply modus ponens to get that Q is true, if we know that P is true.

Proof. Say $P_1, P_2, \ldots, P_n, P$ and $Q_1, Q_2, \ldots, Q_m, Q$ are arguments. In general, if we can show that

$$(p_1 \wedge p_2 \wedge \cdots \wedge p_n) \to p \equiv (q_1 \wedge q_2 \wedge \cdots \wedge q_m) \to q$$

then $(p_1 \wedge p_2 \wedge \cdots \wedge p_n) \to p$ is a tautology if and only if $(q_1 \wedge q_2 \wedge \cdots \wedge q_m) \to q$ is a tautology. Then by Lemma 1.24, $P_1, P_2, \ldots, P_n, P$ is a valid argument if and only if $Q_1, Q_2, \ldots, Q_m, Q$ is a valid argument. So it is this sufficient method of showing logical equivalences that we will use.

1. Let $h = h_1 \wedge h_2 \wedge \cdots \wedge h_n$. Essentially by exportation/importation,

$$(h \wedge p) \to q \equiv (\neg h \vee \neg p) \vee q \equiv \neg h \vee (\neg p \vee q) \equiv h \to (p \to q).$$

2. Again, let $h = h_1 \wedge h_2 \wedge \cdots \wedge h_n$. Repeating the exportation/importation manipulations, we get

$$(h \wedge \neg q) \to \neg p \equiv (\neg h \vee q) \vee \neg p \equiv \neg h \vee (q \vee \neg p) \equiv \neg h \vee (\neg p \vee q) \equiv h \to (p \to q).$$

3. The n arguments H_i, P for $i = 1, 2, \ldots, n$ are valid if and only if each implication $h_i \to p$ is a tautology. Recall that a conjunction is true if and only each of its components is true. So, all of the $h_i \to p$ are tautologies if and only if

$$(h_1 \to p) \wedge (h_2 \to p) \wedge \cdots \wedge (h_n \to p)$$

is a tautology. By the natural generalization of the equivalence

$$(p \to r) \wedge (q \to r) \equiv (p \vee q) \to r$$

to more conjunctions, we get the equivalence

$$(h_1 \to p) \wedge (h_2 \to p) \wedge \cdots \wedge (h_n \to p) \equiv (h_1 \vee h_2 \vee \cdots \vee h_n) \to p.$$

4. As before, let $h = h_1 \wedge h_2 \wedge \cdots \wedge h_n$. Then

$$(h \wedge \neg p) \to \bot \equiv \neg(h \wedge \neg p) \vee \bot \equiv \neg h \vee p \vee \bot \equiv \neg h \vee p \equiv h \to p.$$

For the first special case, the two stated arguments are valid if and only if $(h \wedge \neg p) \to q$ and $(h \wedge \neg p) \to \neg q$ are both tautologies, which is true if and only if

$$((h \wedge \neg p) \to q) \wedge ((h \wedge \neg p) \to \neg q)$$

is a tautology. This is a tautology if and only if its logically equivalent counterpart

$$(h \wedge \neg p) \to (q \wedge \neg q) \equiv (h \wedge \neg p) \to \bot \equiv h \to p$$

is a tautology, which is the first case that we proved. Thus, we can take $\bot$ to be the contradiction $Q \vee \neg Q$ in this case for some proposition Q.

For the special case within this special case, taking Q to be P yields

$$\begin{aligned} h \to p &\equiv ((h \wedge \neg p) \to p) \wedge ((h \wedge \neg p) \to \neg p) \\ &\equiv ((h \wedge \neg p) \to p) \wedge (\neg(h \wedge \neg p) \vee \neg p) \\ &\equiv ((h \wedge \neg p) \to p) \wedge (\neg h \vee p \vee \neg p) \\ &\equiv ((h \wedge \neg p) \to p) \wedge (\neg h \vee \top) \\ &\equiv ((h \wedge \neg p) \to p) \wedge \top \\ &\equiv (h \wedge \neg p) \to p. \end{aligned}$$

There is a dramatic passage about proof by contradiction in Hardy's *A Mathematician's Apology.* He says that contradiction "is one of a mathematician's finest weapons. It is a far finer gambit than any chess gambit: a chess player may offer the sacrifice of a pawn or even a piece, but a mathematician offers *the game.*"

De Morgan's laws are useful in contrapositive or contradiction proofs, as both methods involve negations. □

Definition 1.30. When we say that we are making an assumption **without loss of generality**, what we mean is that we are doing proof by casework, but only some of the cases need to be considered because the other cases follow in a similar way or the cases we are considering can be used to prove the remaining cases.

Example. An example of assumption without loss of generality is if an equation is symmetric in a pair of variables, in which case we can assume one is greater than or equal to the other (Theorem 5.38).

Theorem 1.31. Biconditional propositions, like $P \leftrightarrow Q$, may be proven in several ways using certain combinations of converses, contrapositives, and inverses. More generally, we often want to prove that a collection $P_1, P_2, \ldots, P_n$ (not just a pair) of propositions are pairwise equivalent, meaning $P_i \to P_j$ is true for each ordered pair i, j. This becomes a formidable task as n increases, even if we remain organized. One method of organization is a cyclic proof: prove that $P_1 \to P_2$, and $P_2 \to P_3$, and so on until $P_{n-1} \to P_n$, and finally $P_n \to P_1$ is true. This is not always easy or even possible. A second general option is to pick one particular P_i that is "central" in the sense that we can show that $P_i \leftrightarrow P_j$ for all other indices j.
More generally, if we represent the propositions as distinct dots on a sheet of paper and the implications that we can prove between ordered pair of dots as arrows, it must be possible to start at any node and reach any other node by following the directions of some arrows. This can be done in many ways as n increases. In a proof, what we desire is have certainty that we can traverse the nodes through arrows in this way. Typically, we pick arrows to prove that are relatively convenient (or at least feasible) to prove.

Proof. Such a collection of nodes and arrows represents a collection of pairwise equivalent propositions by transitivity of implication. For example, $P_1 \to P_2, P_2 \to P_3, P_1 \to P_3$ is a valid argument. □

Theorem 1.32 (Implicit biconditionality)**.** Let $P_1, P_2, \ldots, P_n$ be propositions that are collectively exhaustive, meaning $P_1 \vee P_2 \vee \cdots \vee P_n$ is true, or, equivalently, that at least one of the P_i is true. Let $Q_1, Q_2, \ldots, Q_n$ be propositions that are mutually exclusive, meaning $Q_i \wedge Q_j$ is false for every ordered pair i, j such that $i \neq j$. Suppose $P_i \to Q_i$ is true for each $i = 1, 2, \ldots, n$. Then the P_k are mutually exclusive, the Q_k are collectively exhaustive, and $Q_i \to P_i$ is true for each i. This is not merely an abstract result; it actually comes up from time to time and saves us from having to prove some implications, as this result proves them automatically.

Proof. We strongly recommend drawing out these proofs in terms of nodes and arrows.

1. Suppose, for contradiction, that P_i and P_j both hold for unequal i and j. By modus ponens on $P_i \to Q_i$ and $P_j \to Q_j$, both Q_i and Q_j hold. This means $Q_i \wedge Q_j$ holds, which contradicts the mutual exclusivity of the Q_k. Thus, P_i and P_j cannot hold simultaneously, making the P_k mutually exclusive.

2. Since the P_k are collectively exhaustive, some P_i must hold. By modus ponens on $P_i \to Q_i$, we know that Q_i holds as well. Thus, the Q_k are collectively exhaustive.

3. We want to show that for each i, it holds that $Q_i \to P_i$. As in a direct proof, suppose Q_i holds. Since the P_k are collectively exhaustive, at least one of them must hold (in fact, we proved that exactly one holds). If it is P_j that holds for some $j \neq i$, then modus ponens on $P_j \to Q_j$ would yield that Q_j holds, contradicting the mutual exclusivity of the Q_k. So the only possibility is for P_i to hold. Thus, $Q_i \to P_i$ is true.

□

Definition 1.33. There are notions of relative strength and weakness between propositions.

- Suppose two arguments T_1 and T_2 have the same conclusion but different hypotheses H_1 and H_2, respectively. Then H_1 is **weaker** than H_2 if H_2, H_1 is a valid argument. For example, they might both be conjunctions and H_2 has all of the components of H_1 and more.

- Suppose two arguments T_1 and T_2 have the same hypotheses but different conclusions C_1 and C_2, respectively. Then C_1 is **stronger** than C_2 if C_1, C_2 is a valid argument. For cxample, they might both be conjunctions and C_1 has all of the components of C_2 and more.

In either case, or a combined scenario where T_1 has both a weaker hypothesis and a stronger conclusion than T_2, we say that T_1 is a **stronger argument** than T_2 and that T_2 is a **weaker argument** that T_1. On this note, sometimes we do not need the full strength of a theorem, in which case we can say that we have applied it **weakly**; for example, only one direction of a bidirectional result must be needed or the theorem might have conclusions that we do not need. In the other direction, we might have superfluous facts at our disposal which do not need to be fed into the theorem, but there is no well-used terminology for this.

Problem 1.34. Prove that, if the valid argument T_1 is stronger than the argument T_2, then T_2 is a valid argument.

Our advice on proofs of propositions is to not focus on formalities when practicing mathematics. The rules that logicians have apparently invented would not be worth the paper on which they are written if they were not inevitable from the nature of our intuitive reasoning and how reality works. Moreover, we emphasize that practicing mathematicians do not write formal proofs where every step is shown. Rather, they write informal proofs that emphasize the cleverest steps, while glossing over more standard aspects of the proof. Certainly, the names of axioms and rules of inference are rarely mentioned.

1.3 Predicates

Definition 1.35. Propositions alone are not enough to capture all of the kinds of mathematical statements that we would like to make. This is because the most interesting mathematical statements are not static like propositions, but are dynamic in the sense that they involve variables based on which the truth value can change. A statement that involves one or more variables that represent objects is called a **predicate** if fixing all variables in any acceptable way produces a proposition. We denote a predicate as $P(x_1, x_2, \ldots, x_n)$, where the x_i are **variables** and P is the **propositional function** or predicate. Note that we are not allowing propositional variables here, but rather variables that represent mathematical objects like numbers.

Example. The statement $x < 0$ is a predicate because its truth value depends on the variable x. For $x = -1$, it is true that $x < 0$, whereas $x < 0$ is false for $x = 1$.

Definition 1.36. In mathematical writing, we wish to make two basic kinds of assertions about predicates such that the assertions have a fixed true or false value:

- **Universal quantification** means all possible values of the variables lead to true propositions. This is indicated by $\forall x, P(x)$. The **universal quantifier** $\forall$ is read as "for all" or "for each" or "for any" or "given any."
- **Existential quantification** means some particular values of the variables lead to a true proposition. This is indicated by $\exists x, P(x)$. The **existential quantifier** $\exists$ is read as "there exists" or "for some."

A **quantified predicate** is a predicate, all of whose variables are quantified; indeed, in many scenarios, one quantifier is insufficiently descriptive. We interpret a block of quantifiers by reading them from left to right, meaning outside to inside, unlike functions which are applied right to left (so, inside to outside). In terms of the order of operations, we define $\forall$ and $\exists$ to work on a higher level than all logical operators. When we say that two quantified predicates are "equal" and use the $=$ sign, we mean that either both are true or neither is true.

Example. A sentence that uses two quantifiers is "For each dog, there exists a cat, such that the dog chases the cat." We must be careful because it is not always acceptable to swap two quantifiers in a sequence. For example, the meaning of the sentence "There exists a cat such that, for every dog, the dog chases the cat." is different from the meaning of the first sentence. The cat would be very tired from running in the second scenario! The general principle is elucidated in Theorem 1.40.

Theorem 1.37. There are some fairly straightforward inference rules for quantified predicates:

- Universal instantiation: $\forall x, P(x)$ leads to $P(c)$ for any constant c
- Universal generalization: if we know that $P(c)$ holds for any arbitrary constant c, then $\forall x, P(x)$
- Existential instantiation: $\exists x, P(x)$ leads to $P(c)$ for some particular constant c
- Existential generalization: if we know that $P(c)$ holds for a particular constant c, then $\exists x, P(x)$

These are important practical tools in proof-writing as they allow us to jump from quantified predicates to particular propositions and vice versa, thereby allowing us to utilize the rules of inference for propositions in proofs of quantified predicates.

Definition 1.38. By drawing a truth table, we can prove that $\neg P, P \to Q$ is a valid argument. So, if P is false, then $P \to Q$ is true regardless of the truth value of Q. If such an argument is sound, that is if P is false, then we say that the implication $P \to Q$ is **vacuously true**.

Example. Vacuous truth rarely comes up in practice, but there are some scenarios that invite it. For example, in a quantified predicate like $\forall x, (P(x) \to Q(x))$, some of the x might make $P(x)$ false. An example of this is $\forall x, (x \geq 1 \to x \geq 0)$, where x is restricted to the set of integers. If $x < 1$, then the implication is vacuously true; if $x \geq 1$, the implication is genuinely true. In fact, a universally quantified predicate like this might admit no non-vacuous truths whatsoever. Separately, for those familiar with the language of sets, an example is $\forall x, (x \in S \to P(x))$, where S is the empty set. Since the empty set has no elements, $x \in S$ is always false. Vacuous truths can also sometimes be relevant in the base case of proof by induction, which is a method of proof that is the topic of Section 2.3.

Definition 1.39. In proving a quantified predicate of the form $\exists x, P(x)$, we need to find only one example of x for which $P(x)$ is true. This is called a **proof of existence**. There is some additional terminology related to existence proofs:

- If the proof explicitly finds a formula for the object x, then the proof is said to be **constructive**. Otherwise, the proof is called **non-constructive**. Non-constructive proofs assert the existence of x without constructing x. Some possible reasons for not constructing x are that it is not needed, it is more difficult to do so, it is outside of the reach of known methods, or it is impossible.
- As in Definition 1.44, the result might require **uniqueness** of x, meaning we want to prove that $\exists! x, P(x)$. So there is no other example. Proofs of uniqueness often involve deriving a contradiction by assuming that a second solution exists.

Example. An example of a proof of existence is a disproof (this means the proof of the negation) of a universally quantified predicate like $\forall x, P(x)$. A particular x such that $\exists x, \neg P(x)$ is said to be a **counterexample** to $\forall x, P(x)$. Pro tip: it is a good idea to check a theorem against edge cases that are convenient to compute in order to check for easy counterexamples. On the other hand, the disproof of $\exists x, P(x)$ is a proof of $\forall x, \neg P(x)$, so *every* x has to fail to satisfy $P(x)$.

Theorem 1.40. If some quantifiers are in a sequence in a quantified predicate, then the quantifiers along with their corresponding quantified variables may be swapped if and only if the quantifiers are of the same type (both universal or both existential) and they are not separated by a quantifier of the opposite type. For example,

$$\forall x, \forall y, \exists z, P(x, y, z) = \forall y, \forall x, \exists z, P(x, y, z).$$

However, we cannot swap $\forall x$ and $\exists y$ in $\forall x, \exists y, \forall z, P(x, y, z)$ because their quantifiers are of different types, and we cannot swap $\forall x$ and $\forall z$ there either because they are separated by $\exists y$.

Proof. Here is an informal proof of why swapping is allowed under the stated circumstances. It is because a sequence like

$$\forall x_1, \forall x_2, \ldots, \forall x_n, P(x_1, x_2, \ldots, x_n)$$

may be interpreted as

$$\forall \{x_1, x_2, \ldots, x_n\}, P(x_1, x_2, \ldots, x_n)$$

where the variables become a single block variable where order does not matter, and similarly for $\exists$. The order does not matter because the symbols are distinct, so they retain their place in the predicate P. □

Definition 1.41. If quantifiers are placed haphazardly (instead of in a single sequence preceding the predicate), like in $(\forall x, P(x)) \wedge (\exists y, Q(y))$, the best policy is to not make swaps. Though we will not prove it, a theorem of logic states that every quantified predicate, no matter how the quantifiers are scattered, can be written with all of the quantified variables preceding the predicate. This is called **prenex normal form**. After conversion to prenex normal form, it would be possible to see if a swap is legal, as described in Theorem 1.40.

Problem 1.42. It holds (and the reader should check intuitively) that

$$\forall x, (P(x) \wedge Q(x)) = (\forall x, P(x)) \wedge (\forall x, Q(x)),$$
$$\exists x, (P(x) \vee Q(x)) = (\exists x, P(x)) \vee (\exists x, Q(x)).$$

So, $\forall$ distributes over conjunction and $\exists$ distributes over disjunction. Prove, by finding examples, that $\forall$ does not distribute over disjunction and $\exists$ does not distribute over conjunction.

Theorem 1.43 (De Morgan's laws for quantifiers). Universally or existentially quantified predicates may be negated using the following laws:

$$\neg(\forall x, P(x)) = \exists x, \neg P(x),$$
$$\neg(\exists x, P(x)) = \forall x, \neg P(x).$$

In this sense, $\forall$ and $\exists$ are opposites. Negating a predicate with multiple quantifiers follows the same rules: flip each $\forall$ into $\exists$ and each $\exists$ into $\forall$, and negate the underlying predicates.

Proof. This should make sense intuitively, since "there does not exist" means the opposite of the proposition is true in every case, and "not for all" means an example of the opposite of the proposition exists. As for the generalization to multiple quantifiers, we can prove it from the case of one quantifier by converting to prenex normal form and interpreting one quantifier at a time from left to right. For example,

$$\neg(\forall x, \exists y, P(x,y)) = \neg(\forall x, (\exists y, P(x,y))) = \exists x, \neg(\exists y, P(x,y)) = \exists x, \forall y, \neg P(x,y).$$

In a sense, we let $\exists y, P(x,y)$ be $Q(x)$ and proceed from there. □

Definition 1.44. Sometimes, existential quantification is not enough. We also want to assert the **uniqueness** of the object that exists. In that case, we write $\exists! x, P(x)$ to mean

$$\exists x, (P(x) \wedge \forall y, (y \neq x \to \neg P(y))).$$

Equivalently, by contrapositive it also means

$$\exists x, (P(x) \wedge \forall y, (P(y) \to y = x)).$$

Note that the word "**the**" implicitly asserts both existence and uniqueness in mathematics. Similarly, sometimes we want to say that there exist some finite number of satisfactory x, say n of them. In that case, there is no standard notation for it, but it is good practice to clarify in words if we mean that there exist **exactly** or **at least** n examples of x.

Definition 1.45. In making quantified assertions, we often wish to restrict the domain of the variables over which the assertion is being made. Let $R(x)$ be a predicate and $P(x)$ be either a predicate in only the variable x or a quantified predicate where every variable other than x has been quantified. Then we denote and define **bounded** variables that have universal or existential quantifiers, respectively, as

$$\begin{aligned}(\forall x, R(x)), P(x) &= \forall x, (R(x) \to P(x)),\\ (\exists x, R(x)), P(x) &= \exists x, (R(x) \wedge P(x)),\end{aligned}$$

where the = sign means that it is a definition. These definitions intuitively make sense. The predicate $R(x)$ usually involves a set relation, like $\in$ or $\subseteq$, or it is an inequality symbol, such as $<$ or $\geq$.

Example. If there are multiple quantifiers, like in $(\forall x, R(x)), (\exists y, S(y)), P(x,y)$, then we let $Q(x)$ stand for the inner quantified predicate $(\exists y, S(y)), P(x,y)$ and apply the definition of a bounded quantifier to get

$$\begin{aligned}(\forall x, R(x)), (\exists y, S(y)), P(x,y) &= (\forall x, R(x)), Q(x)\\ &= \forall x, (R(x) \to Q(x))\\ &= \forall x, (R(x) \to (\exists y, S(y)), P(x,y))\\ &= \forall x, (R(x) \to \exists y, (S(y) \wedge P(x,y))).\end{aligned}$$

In this way, we can always work from the outside to the inside, from left to right, by using parentheses and substitutions. However, we caution the reader to not unduly follow this overly formal route because it becomes a nightmare as the number of bounded variables increases. While removing the bounds like this may be useful for understanding the nature of a predicate with bounded variables, it is better to naturally interpret bounded variables in terms of restricting the domain of the variable.

Problem 1.46. Prove that the bounding part of a bounded quantification does not change under negation. In symbols, this would mean

$$\begin{aligned}\neg((\forall x, R(x)), P(x)) &= (\exists x, R(x)), \neg P(x),\\ \neg((\exists x, R(x)), P(x)) &= (\forall x, R(x)), \neg P(x).\end{aligned}$$

Thus, we may negate quantified predicates that have bounded variables by pretending that the bounds are not there and then pop them back in unchanged.

Chapter 2

Sets and Maps

"From the paradise, that Cantor created for us, no-one shall be able to expel us."

– *David Hilbert, Über das Unendliche*

"The axiom of choice is obviously true, the well-ordering principle obviously false, and who can tell about Zorn's lemma?"

– *Jerry Bona*

Sets are collections of elements. One way of formulating all objects in the mathematical universe is to boil them down to sets, and, ultimately, combinations of the empty set. We will look at some of the axioms that sets must satisfy and standard methods of constructing sets. Then we will study functions. A function is a machine that takes inputs from a set and, for each input, provides exactly one output in another set. There are numerous general properties of functions that we will see here in preparation for upcoming material. We will end with a proof method that sets allow us to describe: proof by induction and its variants, which are the well-ordering principle and complete induction.

2.1 Sets

The notation of standard logical operators will be used here. We provided a detailed, but practical, viewpoint of logic in Chapter 1.

Definition 2.1. A **set** is a collection of elements, where the elements are not ordered (meaning there is no "first" or "second" or otherwise ranked element) and there are no repeated elements. If it is physically possible on paper, the elements of a set can be written out by enclosing its elements in any order from left to right within curly brackets, like $\{\ldots\}$. We denote that x is an **element of** the set S by writing $x \in S$; if x is not an element of S, then we denote it as $x \notin S$. The notions of "element" and "element of" are undefined, but we have to start somewhere. The set with no elements is called the **empty set** and is denoted by $\varnothing$. A set that is not the empty set is called **non-empty**. Sets S and T are said to be **equal** if

$$\forall x, (x \in S \iff x \in T).$$

Equality of two sets S and T is written using an **equals sign** $=$, like $S = T$. Equal sets may be substituted in for each other in math. A **singleton** is a set with a single element, that is a set of the form $\{x\}$, where x is the only element. An **unordered pair** is a set with exactly two elements like $\{x, y\}$.

Example. In the modern mathematical universe, all objects (e.g. numbers, spaces, and functions) can be expressed in terms of sets, though most of these constructions are too advanced for us, and often too complicated for practical usage. The set of integers is denoted by

$$\mathbb{Z} = \{\ldots, -2, -1, -0, 1, 2, \ldots\},$$

the set of positive integers by

$$\mathbb{Z}_+ = \{1, 2, 3, \ldots\},$$

and the set of non-negative integers by

$$\mathbb{Z}_{\geq 0} = \{0, 1, 2, 3, \ldots\}.$$

In some contexts, the notation $\mathbb{N}$ and the term "natural numbers" is used. We have intentionally avoided this notation and terminology because there is no universal agreement on whether the naturals should include 0. We will study integers in some detail in Chapter 3 and in great detail when we study number theory.

Definition 2.2. A set X is said to be a **subset** of a set Y if

$$\forall x, (x \in X \implies x \in Y),$$

and this is denoted by $X \subseteq Y$. A **proper subset** X of a set Y is a subset other than Y itself. In other words, there exists an element (not necessarily a unique element) that is inside Y but outside X. The assertion that X is a proper subset of Y is denoted by $X \subsetneq Y$. If $X \subseteq Y$, then Y is called a **superset** of X. If we wish to emphasize the superset relation instead of the subset relation, then is acceptable to reverse the notation and write $Y \supseteq X$ instead of $X \subseteq Y$, and $Y \supsetneq X$ instead of $X \subsetneq Y$. We will avoid the notation $\subset$ and $\supset$ to prevent ambiguity; $\subseteq$ and $\supseteq$ denote that equality is possible (but not necessary), whereas $\subsetneq$ and $\supsetneq$ denote that equality is impossible. To denote that X is not a subset of Y we write $X \nsubseteq Y$, and $Y \nsupseteq X$ denotes that Y is not a superset of X.

Example. For every set S, it is vacuously true that $\varnothing \subseteq S$. This includes the fact that $\varnothing$ is a subset of itself. In fact, $\varnothing$ is a proper subset of every non-empty set, yet, in contrast, the only subset of $\varnothing$ is $\varnothing$, so the empty set has no proper subsets.

Theorem 2.3. The subset relation on the subsets of a particular set produces a partial order, which we will define more generally in Definition 5.45. What this means is that for all subsets X, Y, Z of a particular set, the following properties hold:

1. Reflexivity: $X \subseteq X$
2. Antisymmetry: if $X \subseteq Y$ and $Y \subseteq X$, then $X = Y$. Reflexivity gives the converse, thanks to the substitution property of equality (Theorem 2.21).
3. Transitivity: if $X \subseteq Y$ and $Y \subseteq Z$, then $X \subseteq Z$

All of these properties follow from the definition of the subset relation. We leave the verification of the details to the reader.

Definition 2.4. One way of constructing sets is to use **set builder notation**. This allows us to denote and define sets, like

$$\{x \in S : p(x)\} = \{x : x \in S \wedge p(x)\},$$

where $p(x)$ is a predicate (meaning, a property) in the variable x. The colon is read as "such that" and could be replaced by a vertical bar $|$. So we are defining $\{x \in S : p(x)\}$ to be the subset of S such that those are all the elements $x \in S$ that make $p(x)$ true.

One way in which we might wish to extend set builder notation is to say that "this is the set of *all* possible sets that have a certain property." So there is a certain condition and we want the power to summon into a set *every* set that satisfies this predicate p, not just a subset of some fixed set S. Such a strong ability quickly leads to trouble, as shown by Russell's paradox below.

Theorem 2.5 (Russell's paradox). Suppose it is possible to summon all sets x in the mathematical universe that satisfy any given predicate $p(x)$ in the variable x; this is called "unrestricted comprehension." Then we can use this power to deduce a contradiction. For the same reason, there can be no set that contains all sets as elements.

Proof. Suppose, for contradiction, that unrestricted comprehension is acceptable. This allows us to define the set

$$S = \{x : x \notin x\}.$$

Then there are two possibilities: $S \in S$ or $S \notin S$.

- If $S \in S$, then $S \notin S$ by the definition of S. This contradicts the premise of the case in which we are working, that is $S \in S$.
- If $S \notin S$, then $\neg(S \notin S)$ must be true by the definition of S, which is equivalent to $S \in S$. Again, we have contradicted the premise of the case in which we are working, that is $S \notin S$.

In either case, we have a contradiction. Thus, we cannot allow unrestricted comprehension, and we choose the safe-seeming option of summoning only subsets of known sets.
Now suppose there exists a set V that contains all sets. Then we can still define

$$T = \{x \in V : x \notin x\}.$$

Once again, $T \in T$ leads to $T \notin T$, and $T \notin T$ leads to $T \in T$. Since Russell's paradox takes hold again, there can be no such object as the "set of all sets." □

Corollary 2.6. Due to Russell's paradox, we choose to restrict the ability to build sets according to satisfying a certain predicate (in set builder notation) by only allowing the ability to carve out a subset of a set, and not carve out a part of the entire mathematical universe. That is, $\{x \in S : p(x)\}$ is allowed but $\{x : p(x)\}$ is not allowed. Moreover, since the issue of Russell's paradox remains if there is a set whose elements are all possible sets, we must also stipulate that there is no "set of all sets."

Definition 2.7. Given sets X and Y, some common constructions of sets, called set operations, are:

- **Union**: $X \cup Y = \{x : x \in X \vee x \in Y\}$

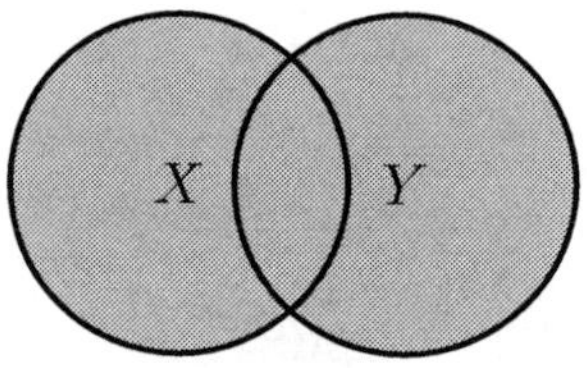

$X \cup Y$

- **Intersection**: $X \cap Y = \{x : x \in X \wedge x \in Y\}$

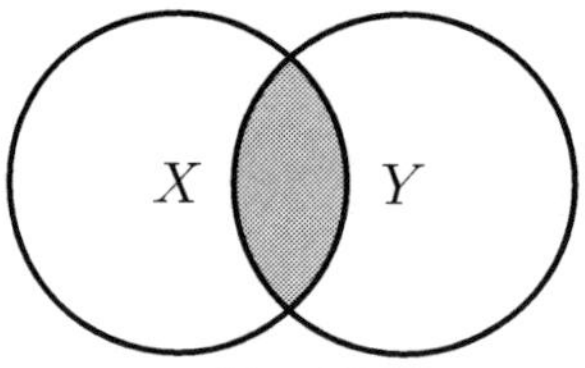

$X \cap Y$

- **Difference**: Also called the **relative complement** of X excluding Y, this is

$$X \backslash Y = X - Y = \{x : x \in X \wedge x \notin Y\}.$$

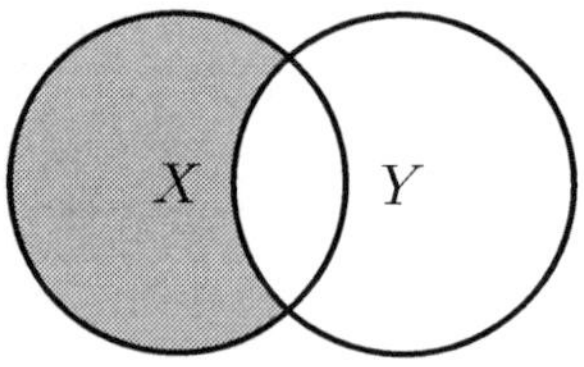

$X \backslash Y$

- **Complement**: If all relevant sets are subsets of some "universal" set $\mathcal{U}$, then we define

$$\overline{X} = X^c = \{x : x \in \mathcal{U} \wedge x \notin X\} = \mathcal{U} \backslash X.$$

Any mention of a complement implies that we are working in an overarching set $\mathcal{U}$. A very useful application of the complement is to write the difference as $X \backslash Y = X \cap (Y^c)$.

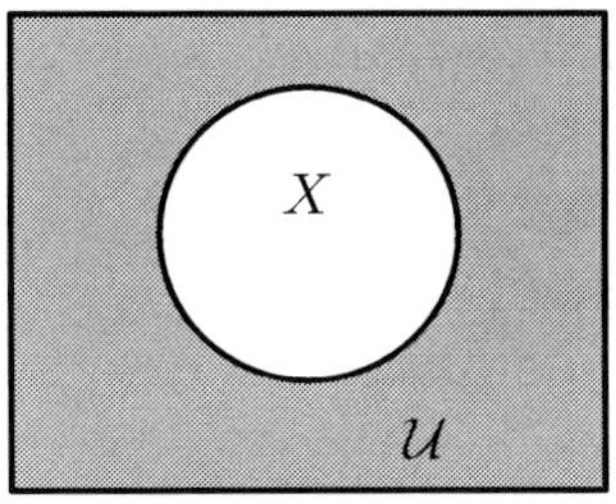

X^c

- **Symmetric difference**: Using difference twice, we define

$$X \oplus Y = (X \backslash Y) \cup (Y \backslash X) = (X \cup Y) \backslash (X \cap Y)$$

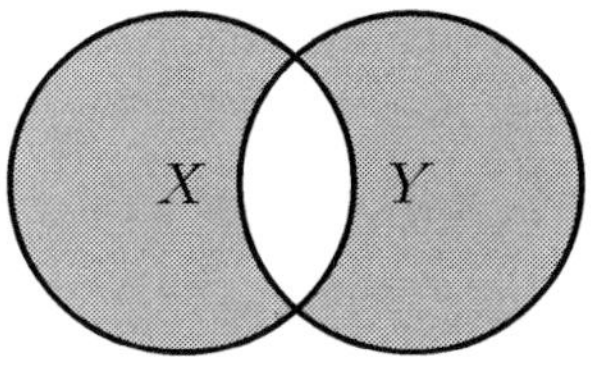

$X \oplus Y$

There is no universally accepted order of operations on sets for the most part. However, complement takes precedence over everything, other than brackets (this is similar to exponents for numbers). So we can write $X \cap Y^c$ instead of $X \cap (Y^c)$.

Lemma 2.8. If X, Y, Z are sets, then:

1. $X \subseteq Z$ and $Y \subseteq Z$, if and only if $X \cup Y \subseteq Z$.
2. $Z \subseteq X$ and $Z \subseteq Y$, if and only if $Z \subseteq X \cap Y$.

These should be clear from the definition of the subset relation.

Theorem 2.9. It is easy to see that for any set X, taking the complement twice yields $(X^c)^c$. Also,

$$\varnothing^c = \mathcal{U} \text{ and } \mathcal{U}^c = \varnothing.$$

The following table provides more such equalities. These identities can be useful in simplifying expressions involving sets. Let X, Y, Z be sets, $\varnothing$ denote the empty set as usual, and, if relevant, let $\mathcal{U}$ denote a universal set in which all work is being done. Then:

Identity	Union	Intersection
Complement	$X \cup X^c = \mathcal{U}$	$X \cap X^c = \varnothing$
Annihilation	$X \cup \mathcal{U} = \mathcal{U}$	$X \cap \varnothing = \varnothing$
Identity	$X \cup \varnothing = X$	$X \cap \mathcal{U} = X$
Idempotent	$X \cup X = X$	$X \cap X = X$
Absorption	$X \cup (X \cap Y) = X$	$X \cap (X \cup Y) = X$
Commutative	$X \cup Y = Y \cup X$	$X \cap Y = Y \cap X$
Associative	$X \cup (Y \cup Z) = (X \cup Y) \cup Z$	$X \cap (Y \cap Z) = (X \cap Y) \cap Z$
Distributive	$X \cup (Y \cap Z) = (X \cup Y) \cap (X \cup Z)$	$X \cap (Y \cup Z) = (X \cap Y) \cup (X \cap Z)$
De Morgan's	$(X \cup Y)^c = X^c \cap Y^c$	$(X \cap Y)^c = X^c \cup Y^c$

The associative laws allow us to write a sequence of unions or a sequence of intersections without parentheses, but not if unions and intersections are mixed. The commutative laws allow us to write a sequence of unions in any order, and similarly for intersections, but again not if they are mixed. When unions and intersections both appear, other equivalences like the absorption laws and distributive laws may be helpful. Unlike set union and intersection, set difference is neither commutative not associative, meaning it is not necessarily true that $X \backslash Y = Y \backslash X$. However, the symmetric difference of sets is both commutative and associative.

Set identities can be proven by using other identities, or logical definitions and logical equivalences of the predicates involved, or via the antisymmetry of the subset relation. We will not cover any of these proofs because they are too cumbersome and space-consuming.

Corollary 2.10. By repeated applications of de Morgan's laws, these two laws may be extended to the union of many sets or the intersection of many sets:

$$(S_1 \cup S_2 \cup \cdots \cup S_n)^c \equiv S_1^c \cap S_2^c \cap \cdots \cap S_n^c,$$
$$(S_1 \cap S_2 \cap \cdots \cap S_n)^c \equiv S_1^c \cup S_2^c \cup \cdots \cup S_n^c.$$

Thanks to de Morgan's laws, it is often more convenient to work with complements than to take a difference with a universal set.

Problem 2.11. Let A and B be sets such that their union is a universal set $\mathcal{U}$ and their intersection is the empty set $\varnothing$. Prove that $A^c = B$.

Theorem 2.12. Without getting into the names or technical statements of the axioms involved, here are a few properties satisfied by sets:

- No set can be an element of itself.
- For any two sets X, Y, there is a "pairing" set $\{X, Y\}$ that contains just these sets as its two elements. This can be used repeatedly to produce a set of n sets. Alternatively, we can take $X = Y$ to get that there is a singleton set $\{X\}$ (which is different from X).
- If S is a set of sets, then there is a set that is the union of the elements of S. For example,

$$\bigcup\{\{1, 2, 3\}, \{2, 3, 4\}, \{6, 7\}\} = \{1, 2, 3, 4, 6, 7\}.$$

Definition 2.13. Two sets A and B are said to be **disjoint** if they have no shared elements, meaning $A \cap B = \varnothing$. A non-empty set S of sets is said to be **pairwise disjoint** if each pair of distinct sets $A, B \in S$ is disjoint; by distinct, we mean that $A \neq B$. We will not classify the empty set as pairwise disjoint so that there will never be a need to clarify that a pairwise disjoint set is non-empty; one could easily define otherwise, if desired. The union of a pairwise disjoint set of sets is is called a **disjoint union**. If we wanted to emphasize the fact that a union is specifically a disjoint union, we would use the rigid $\sqcup$ symbol instead of the rounded $\cup$ symbol.

Example. Since $\varnothing \cap \varnothing = \varnothing$, our definition says that $\varnothing$ and $\varnothing$ are disjoint. Secondly, all sets whose only element is one set are pairwise disjoint. Entertainingly, this means that, although we have defined the empty set $\varnothing$ to not be pairwise disjoint, the set containing only the empty set $\{\varnothing\}$ is pairwise disjoint.

Definition 2.14. An **ordered pair** is similar to a set of two elements, except in two regards: order matters, meaning there is a "first" element and a "second" element, and the two elements are allowed to be the same. If the first element is a and the second element is b, then the ordered pair is written as (a, b). According to Kuratowski, it is possible to define this ordered pair using sets as $\{\{a\}, \{a, b\}\}$, but that would not serve any purpose at this level. Two ordered pairs (a, b) and (c, d) are said to be **equal** if they are equal component-wise, meaning $a = c$ and $b = d$.

Definition 2.15. Given an ordered pair of sets (A, B), we denote and define its **Cartesian product** to be the set

$$A \times B = \{(a, b) : a \in A, b \in B\}.$$

So, it is the set of all ordered pairs such that the first entry is from A and the second entry is from B. This definition will be extended to a Cartesian product of n ordered sets in Definition 4.4, after we define lists in Definition 4.1.

Definition 2.16. Let X and Y be sets. A **binary relation** on the ordered pair of sets (X, Y) is a subset of $X \times Y$. If the name of the relation is R, the fact that $(x, y) \in R$ may be denoted by xRy, which we utter as "x is related to y". In a sense, a binary relation assigns a "true" or "false" value to each element of $X \times Y$, with the true ones being the elements of R.

Definition 2.17. If S is a non-empty set, an **equivalence relation** on S is a binary relation $\sim$ on $S \times S$ that satisfies the following three properties for all $a, b, c \in S$:

1. Reflexive: $a \sim a$
2. Symmetric: $a \sim b$ implies $b \sim a$
3. Transitive: $a \sim b$ and $b \sim c$ together imply $a \sim c$

Definition 2.18. Given an equivalence relation $\sim$ on S, and given an element $a \in S$, the **equivalence class** of a is denoted by and defined as

$$[a] = \{b \in S : a \sim b\},$$

which is the set of all $b \in S$ to which a is related. If a is an element of S, and C is an equivalence class of S under $\sim$ such that $a \in C$, then a is said to be a **representative** of C.

Theorem 2.19. An equivalence relation on S splits S into non-empty disjoint equivalence classes whose union is S. Conversely, a partition of a set induces an equivalence relation such that two elements are equivalent to each other if and only if they are in the same component. The following are some useful consequences of this partition idea:

1. Every element of S is in some equivalence class, so the classes cover all of S. On the other hand, no element is in more than one class.
2. All elements in an equivalence class are pairwise equivalent to each other under this equivalence relation.
3. Two elements from different equivalence classes cannot be equivalent to each other.
4. If $a \in S$ is an element of an equivalence class C, then the class of a is equal to C, meaning $[a] = C$. As a result, if $b \in [a]$, then $[a] = [b]$.
5. If $a \sim b$ then $[a] = [b]$. Similarly, any two equivalence classes that share an element are in fact equal. In a sense, a single common element causes the classes to collapse into each other.
6. If $a \nsim b$ then $[a] \cap [b] = \varnothing$. Similarly, any two different equivalence classes are disjoint.

The proofs of these facts are unnecessary at this level, so we will not cover them here. The fact that the set of equivalences classes forms a partition of the set will be proven in Volume 2.

Example. Equivalence relations are among the most important instances of binary relations, though we will see other examples of binary relations, like orders in Section 5.3. The benefit of an equivalence relation is that, even though it does not necessarily mean exact equality, it allows us to zero in on some similar quality of all the objects in the same class, or focus on some property that is preserved between two equivalence classes no matter which of their representatives are chosen. Examples of equivalence relations include the congruence or similarity of triangles and polygons, equipollence of directed line segments in the study of vectors (angles between vectors are preserved no matter which representative directed line segments are chosen), equipotence of sets with regards to cardinality, and congruence of integers in modular arithmetic. As a basic example, Definition 2.20 asserts that equality relations, such as those between sets or numbers or polynomials, are equivalence relations.

Definition 2.20. Equality of objects, such as numbers, in a set can be interpreted as the special "finest" equivalence relation on the set, meaning the equivalence classes are precisely the singletons of the set. So equality holds between two objects if and only if they are the same object. This relation is denoted by the equals sign, like $A = B$, which is called an **equation**. Its negation is denoted by $A \neq B$, which means A and B are different objects. As an equivalence relation, equality satisfies reflexivity, symmetry, and transitivity.

Theorem 2.21 (Substitution property)**.** If a and b are equal objects meaning $a = b$, then as equal objects, we can replace a with b or b with a in an equation or any other mathematical statement without altering the truth value of the statement.

The substitution property should make sense intuitively.

Definition 2.22. There are several definitions and applications of the term, **well-defined**, of which we are aware, all of them being about avoiding ambiguity in interpretation.

- A representation of an object is called well-defined if there is no other such representation. For example, for each integer $b \geq 2$, integers have exactly one terminating representation in base-b.
- A function is well-defined if it produces the same output for a given input, regardless of which of the multiple representations of the input are used. For example, $0.999\ldots$ and 1.0 represent the same rational number, but a function should be independent of the the symbolic representation. An anti-example is the "function" f that takes rational numbers in fraction form and outputs the numerator. This is not a function at all because $f\left(\frac{1}{2}\right) = 1$ and $f\left(\frac{3}{6}\right) = 3$, whereas the inputs $\frac{1}{2}$ and $\frac{3}{6}$ are equal, so the outputs should have also been equal.
- A function that is defined on a set of equivalence classes is well-defined if, even if the function is defined in terms of elements of the equivalence classes, the function does not produce different outputs for different elements of the same class. For example, in the arithmetic of $\mathbb{Z}_n$ that we will mention in Volume 3, the addition
$$[a] + [b] = [a + b]$$
is independent of the choice of representatives a and b of their respective classes.
- Notation is well-defined if we take advantage of some convenient property to relax the notation without introducing ambiguity. For example, the associativity of multiplication allows us to write $a \cdot b \cdot c$ without parentheses, as opposed to specifying $a \cdot (b \cdot c)$ or $(a \cdot b) \cdot c$.

If a representation, function or piece of notation is not well-defined, then we call it **ill-defined**.

2.2 Functions

Definition 2.23. A **variable** is an unknown or unspecified mathematical object, and may be represented using a symbol, such as x. A variable derives its name from the fact that it is allowed to *vary* over a certain domain (which is often the real numbers).

Definition 2.24. Let X and Y be non-empty sets. A binary relation R on (X, Y) is called **functional** if xRy_1 and xRy_2 together imply that $y_1 = y_2$; so each $x \in X$ is related to *at most* one element of Y. A relation R is called **serial** if
$$\forall x \in X, \exists y \in Y, xRy,$$
so each $x \in X$ is related to *at least* one element of Y. A **function** from X to Y is a binary relation on (X, Y) that is both functional and serial. Less formally, a function $f : X \to Y$ takes inputs from X, and for each $x \in X$, produces exactly one output $y \in Y$, where the output (otherwise known as the **image** of x under f) is denoted by $f(x)$. We usually refer to the function itself as f, though the function can also be denoted by $f(x)$ if x is a general variable that does not represent any particular element of X. In the equation $y = f(x)$, there is exactly one y-value assigned to each x-value, so x is called the **independent variable** and y is called the **dependent variable**. While notation such as $f(x) = x + 1$ is usually used to define how outputs are found using inputs, there is also the "maps to" notation, $x \mapsto x + 1$. There are some sets associated with a function:

- The **domain** X of f is the set of all inputs and is denoted by $\mathrm{Dom}(f)$.
- The **range** of f is the set of all outputs that actually occur and is denoted by $\mathrm{Rng}(f)$. It can be proven in set theory that the range of a function is always a set. So, if f is a function with domain X, then the range is the set denoted by and defined as
$$\{f(x) : x \in X\} = \{y : \exists x, (x \in X \land y = f(x))\}.$$

We have defined set-builder notation in Definition 2.4, and it can be modified in the way on the left side of this equation. If the domain is X, the range of f may be denoted by $f(X)$, which is a special case of the more general image notation in Definition 2.26.

- It is not always practical to write down the exact range, so we often prefer to deal with a **codomain** Y instead, which is a (non-unique) set of which the range is a subset. When the notation $f : X \to Y$ is used instead of just f, it means a codomain Y is specified. If a codomain Y is specified, then we can say that f is **Y-valued**.

Two functions are said to be **equal** if they have the same domain and they produce the same outputs for the same elements of the domain. In particular, it is possible for two functions to be equal without having the same codomain. We denote that f and g are equal functions by $f = g$.

Example. The most basic functions are **constant functions**, where $f(x) = c$ for a fixed value c for all x in the domain. After constant functions, the next most basic function with domain X is $\mathrm{Id}_X : X \to X$, which is defined as $\mathrm{Id}_X(x) = x$ for all $x \in X$. This is called the **identity function** on X.

Definition 2.25. If the domain of a function f is **restricted** to a subset I of the domain, then we denote the new function with the restricted domain by $f|_I$.

Definition 2.26. For a set of values I in the domain of f, the image of I under f is denoted by and defined as

$$f(I) = \{f(x) : x \in I\}.$$

This is called the **image** of I under f.

Definition 2.27. If there are functions $f : X \to Y$ and $g : Y \to Z$, then the **composite function** or **composition** $g \circ f : X \to Z$ is defined as

$$(g \circ f)(x) = g(f(x))$$

for all $x \in X$. In this case, it is said that g is **left-composed** with f and f is **right-composed** with g. Composition can occur several times or not at all:

- If a function f is applied n times, then we denote the composition $\underbrace{(f \circ f \circ \cdots \circ f)}_{n \text{ copies of } f}$ by f^n. This is different from exponents and does *not* denote $(f(x))^n$. Note that, in order to compose a function with itself, its range has to be a subset of its domain.
- By definition, f^1 is the same as just f.
- For any function f, the notation f^0 means f is applied zero times, so $f^0(x) = x$ for all $x \in X$. In other words, f^0 is the identity function on the domain of f.

Theorem 2.28. Function composition is associative, meaning, for functions $f : X \to Y, g : Y \to Z, h : Z \to W$, it holds that

$$(h \circ g) \circ f = h \circ (g \circ f).$$

So, it is not ambiguous to simply write it as $h \circ g \circ f$ without any parentheses, and similarly for any other finite sequence of compositions. However, function composition is not commutative in general.

So far, we know what f^n means for $n \geq 0$, from which arises the question of how f^n should be interpreted for negative integers n. If we had a definition of f^{-1}, we could define f^{-n} as n applications of f^{-1}, but we still need a reasonable definition of f^{-1} at first. For non-negative integers n, it is true that $f \circ f^n = f^{n+1} = f^n \circ f$. If we want this property to hold for $n = -1$, then we would require something along the lines of (but not necessarily exactly)

$$f \circ f^{-1} = \mathrm{Id} = f^{-1} \circ f,$$

where Id is an identity function on some set. So, intuitively, f^{-1} should undo or invert the work done by f, and vice versa. As we will see, a function has an inverse if and only if certain circumstances hold. Moreover, we will have to be careful about specifying the domain and codomain of f^{-1}.

Definition 2.29. Let $f : X \to Y$ be a function. Then:

1. A function $g : Y \to X$ is called a **left-inverse** of f if $g \circ f = \mathrm{Id}_X$.
2. A function $g : Y \to X$ is called a **right-inverse** of f if $f \circ g = \mathrm{Id}_Y$.
3. A function $g : Y \to X$ is called an **inverse** of f if g is both a left-inverse and a right-inverse of f. Then g is denoted as f^{-1}. A function f is called **invertible** if it has an inverse g.

Theorem 2.30. Suppose $g : Y \to X$ is a left-inverse of $f : X \to Y$ and $h : Y \to X$ is a right-inverse of f. Then $g = h$, and this common function is an inverse of f.

Proof. Suppose f, g, h are as stated. Then

$$g = g \circ \mathrm{Id}_Y = g \circ (f \circ h) = (g \circ f) \circ h = \mathrm{Id}_X \circ h = h.$$

So $g = h$ is both a left-inverse and a right-inverse of f, making it an inverse of f. □

Problem 2.31. Show that if an inverse of a function exists, then it is unique. In other words, if g and h are inverses of f then $g = h$. This allows us to say "the" inverse instead of "an" inverse. Moreover, find a counterexample to the assertion that, when a left-inverse exists for a function it is unique, and similarly for right-inverses. As a hint, use small finite sets for the counterexamples.

Theorem 2.32. Suppose $f : X \to Y$ and $g : Y \to X$ are functions. Then:

1. g is a left-inverse of f if and only if f is a right-inverse of g.
2. f is invertible with g being its inverse if and only if g is invertible with f being its inverse. This allows us to say that f and g are "inverses of each other" or simply "inverses," instead of specifying which one is the inverse of the other.
3. If f has an inverse, then its inverse is invertible, and the inverse of the inverse of f is f.

Proof. We will prove these one after another, as each will help to prove the next one:

1. By the definition of left and right-inverses, the proposition $g \circ f = \mathrm{Id}_X$ is simultaneously equivalent to saying that g is a left-inverse of f and that f is a right-inverse of g.
2. By the definition of an inverse, f is invertible with g being its inverse if and only if g is a left-inverse and a right-inverse of f. By the previous part, this is equivalent to f being a right-inverse and a left-inverse of g, which is equivalent to g being invertible with f being its inverse.
3. Suppose f has an inverse f^{-1}. By the previous part, this implies that f^{-1} is invertible with f being its inverse. Thus, $(f^{-1})^{-1} = f$.

□

Theorem 2.33. Suppose $f : X \to Y$ and $g : Y \to Z$ are invertible functions. Then $g \circ f$ is invertible and

$$(g \circ f)^{-1} = f^{-1} \circ g^{-1}.$$

More generally, induction (see Section 2.3) can be used to prove that

$$(f_n \circ \cdots \circ f_2 \circ f_1)^{-1} = f_1^{-1} \circ f_2^{-1} \circ \cdots \circ f_n^{-1}.$$

Proof. Let the inverse of f be $f^{-1} : Y \to X$ and let the inverse of g be $g^{-1} : Z \to Y$. Then $f^{-1} \circ g^{-1}$ is a function that maps Z to X. Then

$$(f^{-1} \circ g^{-1}) \circ (g \circ f) = f^{-1} \circ (g^{-1} \circ g) \circ f = f^{-1} \circ \mathrm{Id}_Y \circ f = f^{-1} \circ f = \mathrm{Id}_X,$$
$$(g \circ f) \circ (f^{-1} \circ g^{-1}) = g \circ (f \circ f^{-1}) \circ g^{-1} = g \circ \mathrm{Id}_Y \circ g^{-1} = g \circ g^{-1} = \mathrm{Id}_Z.$$

This proves that $g \circ f$ is invertible with inverse $f^{-1} \circ g^{-1}$. □

Definition 2.34. To find convenient criteria that allow us to establish the existence of a left-inverse, right-inverse, or inverse, we will define the following conditions for a function $f : X \to Y$:

1. f is **injective** or **one-to-one** if, for all x_1 and x_2 in X, $f(x_1) = f(x_2)$ implies $x_1 = x_2$. This means that for each output in the range of f, there is exactly one input in the domain of f that leads to that output. Alternatively formulated, each codomain element has *at most* one domain element mapping to it. By contrapositive, we can also say that different domain elements map to different codomain elements.

2. f is **surjective** or **onto** if, for all $y \in Y$, there exists *at least* one $x \in X$ such that $f(x) = y$. In other words, the range of f is the specified codomain of f.

3. f is **bijective** if it is both injective and surjective.

Problem 2.35. Let $f : X \to Y$ and $g : Y \to Z$ be functions. Prove that, if f, g are injections, then $g \circ f$ is an injection. Similarly, prove that, if f, g are surjections, then $g \circ f$ is a surjection. As a clear implication, this means that if f, g is a bijection, then $g \circ f$ is a bijection.

Example. Though we will not prove it, $f(x) = x^n$ is bijective from $\mathbb{R}$ to $\mathbb{R}$ for any fixed odd integer n. On the other hand, bijectivity of x^n does not hold if n is an even integer, but we regain bijectivity here if we restrict the domain and codomain to $\mathbb{R}_{\geq 0}$.

Definition 2.36. Let $f : X \to Y$ be a function such that $Y \subseteq X$, so that it is possible to compose f with itself. If there exists a positive integer n such that $f^n(x) = x$ for all $x \in X$, then we say that f is **cyclic**. In this case, the smallest positive integer m satisfying $f^m(x) = x$ for all x is called the **order** of f. If the order of a cyclic function f is 2, then f is called an **involution**.

Problem 2.37. Prove that every cyclic function is injective.

Problem 2.38. Show that the function

$$\begin{aligned} f : \mathbb{R}\backslash\{0,1\} &\to \mathbb{R}\backslash\{0,1\}, \\ x &\mapsto \frac{x-1}{x} \end{aligned}$$

is a cyclic function of order 3.

Definition 2.39. If, in set builder notation, we wish to carve out the subset of the image of a set X under a function f such that we use only those elements $x \in X$ that satisfy a predicate $p(x)$, then we get the set

$$\{f(x) : x \in X \wedge p(x)\}.$$

If f is injective, then we may instead use the notation

$$\{f(x) \in f(X) : p(x)\}.$$

We have imposed injectivity here because, as a result of there being a unique input x corresponding to the output $f(x)$, we can test $p(x)$ on the unique x. Otherwise, if there are multiple x's corresponding to $f(x)$, then $p(x)$ might be true for some and false for others, which makes us question whether $f(x)$ should be in the set.

We will need the following definition to complete our discussion of inverses.

Definition 2.40. Let $f : X \to Y$ be a function. For each $y \in Y$, the **preimage** of y under f is denoted by and defined as

$$f^{-1}(y) = \{x \in X : f(x) = y\}.$$

Moreover, for any subset Z of Y, the **preimage** of Z under f is denoted by and defined as

$$f^{-1}(Z) = \bigcup_{z \in Z} f^{-1}(z) = \bigcup \left\{f^{-1}(z) : z \in Z\right\}.$$

Preimages are not mere abstractions for the sake of abstraction. For example, they are of fundamental importance in combinatorics, as we will see in Volume 2. In the theory of real functions, they are related to root-finding, as the following definition shows.

Definition 2.41. If $f : X \to Y$ is a function where Y is a subset of the real (or complex) numbers, then the preimage $f^{-1}(0)$ is called the **zero set** of f and its elements are called the **roots** or **zeros** of f. The elements of $f^{-1}(0)$ are also called the **solutions** of the equation $f(x) = 0$.

The determination of the roots of functions is a fundamental problem in algebra. A part of what makes root-finding an even more frequent activity than one might imagine is that, for any fixed $y \in Y$, the determination of the preimage $f^{-1}(y)$ is the same as finding the roots of $f(x) - y$ or, equivalently, the solutions of $f(x) = y$. We will develop methods of determining the roots of quadratic functions in Theorem 10.5 and Theorem 10.11 and, more generally, for polynomials in Section 11.1.

Definition 2.42. A **choice function** f with domain X, where each element of X is a non-empty set, is a function such that

$$\forall S \in X, f(S) \in S.$$

So a choice function maps each set $S \in X$ to an element of S. The **axiom of choice** states that every set of non-empty sets has a choice function defined on X.

In essence, the axiom of choice says that, given a collection of non-empty boxes, we have the power to choose exactly one item from each box. This makes sense for a finite number of boxes, but the result becomes more controversial as the number of boxes climb to infinity and beyond. There are pathological consequences such as the Banach-Tarski paradox and the well-ordering of the real continuum. We have not discussed these advanced results.

Theorem 2.43. Let $f : X \to Y$ be a function. Then we can determine when f has what kind of inverse according to whether it is injective, surjective or bijective.

1. f is injective if and only if f has a left-inverse.
2. f is surjective if and only if f has a right-inverse.
3. f is bijective if and only if f has an inverse.

An implication of the last property is that the inverse of a bijective function is also bijective.

Proof. We will need the axiom of choice from set theory for the second part of the proof.

1. Suppose f is injective. Define $g : Y \to X$ so that if $y \in f(X)$ then $g(y) = x$ is the unique element of $f^{-1}(y)$, where the uniqueness is an implication of the injectivity of f, and if $y \notin f(X)$ then $g(y) = z$ for some constant $z \in X$, since we are assuming that X is non-empty. Then $g \circ f = \mathrm{Id}_X$ and so g is a left-inverse of f.

   In the other direction, suppose $g : Y \to X$ is a left-inverse of f. If $f(x_1) = f(x_2)$ for some $x_1, x_2 \in X$ then applying g to both sides yields

   $$x_1 = g(f(x_1)) = g(f(x_2)) = x_2,$$

   which proves injectivity.

2. Suppose f is surjective. Then, for each $y \in Y$, the preimage $f^{-1}(y)$ is nonempty. Using the axiom of choice, for each $y \in Y$, we can arbitrarily choose $x \in f^{-1}(y)$ and define $g : Y \to X$ by $g(y) = x$. Then $f \circ g = \mathrm{Id}_Y$ and so g is a right-inverse of f.

   In the other direction, suppose $g : Y \to X$ is a right-inverse of f. To show that f is surjective, we need to show that for each $y \in Y$, there exists an $x \in X$ such that $f(x) = y$. Given $y \in Y$, we let $x = g(y)$. Indeed, applying f to both sides yields

   $$f(x) = f(g(y)) = y.$$

3. An inverse of f exists if and only if it has both a left-inverse and a right-inverse. By the previous two arguments, this is true if and only if f is both injective and surjective, which is equivalent to f being bijective. However, if we wish to bypass the controversial axiom of choice, there exists a proof of this assertion that does not rely on it, as follows.

   In one direction, suppose f is bijective. Injectivity is equivalent to the proposition that, for each $y \in Y$, the preimage $f^{-1}(y)$ has at most one element. Surjectivity is equivalent to the proposition that, for each $y \in Y$, the preimage $f^{-1}(y)$ has at least one element. So bijectivity is equivalent to saying that the preimage $f^{-1}(y)$ of each $y \in Y$ is a singleton. So we construct $g : Y \to X$ by mapping each $y \in Y$ to the unique x such that $f(x) = y$. As desired, it holds that $g \circ f = \mathrm{Id}_X$ and $f \circ g = \mathrm{Id}_Y$.

   In the other direction, suppose f has an inverse g. Then g is a left-inverse and a right-inverse of f. Then we can simply repeat the second direction of each of the previous arguments to show that the existence of a left-inverse implies injectivity and that the existence of a right-inverse implies surjectivity. Neither argument uses the axiom of choice. Thus, f is bijective.

Suppose f is a bijective function whose inverse is g. By Theorem 2.32, f is invertible with g being its inverse if and only if g is invertible with f being its inverse. Thus, by the last property in the above list, g is also bijective. □

Example 2.44. Describe a method of finding the inverse of a bijective function $f : X \to Y$ where X and Y are subsets of $\mathbb{R}$, and where we have an algebraic expression for the definition of f. Apply this method to finding the inverse of the function $f : \mathbb{R} \to \mathbb{R}$ that is defined as $f(x) = (2x + 1)^3$ for all real x.

Solution. Assume that $f : X \to Y$ has an inverse $g : Y \to X$. By the definition of an inverse, $g(f(x)) = x$ for all $x \in X$ and $f(g(y)) = y$ for all $y \in Y$. The former is not too useful in practice. However, if we have an algebraic expression for f, then $f(g(y)) = y$ can be used to get an equation in terms of $g(y)$ that we can solve for $g(y)$. It then remains to be shown that this function g is both a left-inverse and a right-inverse, which can be done by checking that g indeed maps each element of X to some element of Y, and algebraically verifying that $g \circ f = \mathrm{Id}_X$ and $f \circ g = \mathrm{Id}_Y$.
Let us apply this method now. Let $f(x) = (2x + 1)^3$ have domain and codomain $\mathbb{R}$. Assume that f has an inverse $g : \mathbb{R} \to \mathbb{R}$. Then

$$y = f(g(y)) = (2g(y) + 1)^3,$$

which we can solve to get

$$g(y) = \frac{\sqrt[3]{y} - 1}{2}.$$

It can easily be verified that g maps each real number to a real number, and that

$$f \circ g = \mathrm{Id}_{\mathbb{R}} = g \circ f,$$

so g is indeed the inverse of f.
Instead of going through the formal steps that we have stated, an informal method is as follows: We can start with the equation $f(x) = y$ which has y written in terms of x, and manipulate the equation until x is isolated in terms of y. Then the expression in terms of y needs to be shown to be both a left-inverse and a right-inverse of f as in the formal method. Ordinarily, as a final step, we would also replace all original x's with y's and replace all original y's with x's. □

Problem 2.45. Let X and Y be non-empty sets. Show that there exists an injection $f : X \to Y$ if and only if there exists a surjection $g : Y \to X$. You may use the axiom of choice or results that depend on it.

Problem 2.46. Find functions f, g, h, all with domains and codomains equal to $\mathbb{R}$, such that $g \circ f = h \circ f$ and $g \neq h$ simultaneously. Separately, what about simultaneously satisfying $f \circ g = f \circ h$ and $g \neq h$?

Problem 2.47. Find a function $f : \mathbb{R} \to \mathbb{R}$ such that f is injective and not surjective. Separately, find a function $g : \mathbb{R} \to \mathbb{R}$ that is simultaneously surjective and not injective.

Problem 2.48. Let $f : A \to B$ and $g : B \to C$ be functions. Prove that:

1. If $g \circ f$ is injective, then f is injective. Also, find an example of $f : \mathbb{R} \to \mathbb{R}$ and $g : \mathbb{R} \to \mathbb{R}$ such that $g \circ f$ is injective, but g is simultaneously not injective.

2. If $g \circ f$ is surjective, then g is surjective. Also, find an example of $f : \mathbb{R} \to \mathbb{R}$ and $g : \mathbb{R} \to \mathbb{R}$ such that $g \circ f$ is surjective, but f is simultaneously not surjective.

Problem 2.49. Find functions $f : \mathbb{R} \to \mathbb{R}$ and $g : \mathbb{R} \to \mathbb{R}$ such that $g \circ f$ is bijective, but $f \circ g$ is simultaneously not bijective.

The following is a powerful theorem. Due to the difficulty of its proof, we will skip over providing it. However, we will come back to it in Volume 2 to prove a finite version.

Theorem 2.50 (Schröder-Bernstein theorem)**.** If X and Y are non-empty sets and there exists an injection $f : X \to Y$ and an injection $g : Y \to X$, then there exists a bijection $h : X \to Y$. By Problem 2.45 (the solution to which depends on the axiom of choice), each of the following conditions are also individually sufficient for proving the existence of a bijection $h : X \to Y$:

1. There exists an injection $f : X \to Y$ and a surjection $g : X \to Y$.

2. There exists a surjection $f : X \to Y$ and a surjection $g : Y \to X$.

Problem 2.51. Two sets A and B are said to have the "same cardinality" if there exists a bijection $f : A \to B$. Prove that the following sets, strangely enough, all have the same cardinality:

$$[0, 1], (0, 1), \mathbb{R}, (0, \infty), [0, \infty).$$

Interval notation is defined in Definition 5.22. As a hint, Theorem 2.50 will be useful for some of the cases. In Volume 3, we will prove that $\mathbb{R}$ is "uncountable" in the sense that a bijection cannot be set up between $\mathbb{R}$ and $\mathbb{Z}_+$, thereby proving that all of the sets here are also uncountable.

Definition 2.52. The **power set** of a set S is the set of all subsets of S, including the empty set $\varnothing$ and S itself. The power set of S is denoted by $\mathcal{P}(S)$. It is an axiom that, for each set, its power set exists.

Example. The power set of $\{1, 2, 3\}$ is

$$\{\varnothing, \{1\}, \{2\}, \{3\}, \{1, 2\}, \{1, 3\}, \{2, 3\}, \{1, 2, 3\}\}.$$

As an edge case, $\mathcal{P}(\varnothing) = \{\varnothing\}$.

Theorem 2.53 (Cantor's theorem)**.** There is no surjective function from a non-empty set to its power set. However, for each non-empty set, there exists an injective function from the set to its power set.

Proof. The proof will be reminiscent of the proof of Russell's paradox (Theorem 2.5). Let X be a non-empty set. Suppose, for contradiction, that there exists a surjection $f : X \to \mathcal{P}(X)$. Then we define the Cantor diagonal set of f to be the set

$$D = \{x \in X : x \notin f(x)\}.$$

Note that D is a subset of X, so it is an element of $\mathcal{P}(X)$. The assumed surjectivity of f implies that

$$\exists y \in X, f(y) = D.$$

Now there are two possibilities: $y \in D$ or $y \notin D$.

- If $y \in D$, then $y \notin f(y) = D$. This contradicts the premise that $y \in D$.
- If $y \notin D$, then $\neg(y \notin f(y))$, which means $y \in D$. This contradicts the premise that $y \notin D$.

Since we get a contradiction either way, f cannot be surjective. However, $g : X \to \mathcal{P}(X)$, defined by $x \mapsto \{x\}$ is an injection. ☐

Corollary 2.54. What Cantor's theorem implies is that we can repeatedly take power sets of some infinite set X to get sets

$$X, \mathcal{P}(X), \mathcal{P}(\mathcal{P}(X)), \mathcal{P}(\mathcal{P}(\mathcal{P}(X))), \ldots$$

that are "bigger and bigger infinities" in the sense that there can be no surjection from an earlier set to a later set in this sequence of successive power sets, yet a same-sized "copy" of each set is embedded inside each set beyond it via injections.

Proof. Let the sequence of power sets be $p_0, p_1, p_2, p_3, \ldots$ so that p_k is the set where the power set has been applied k times to the original set X. Suppose, for contradiction, that there is a surjection $f : p_m \to p_n$ for some non-negative integers $m < n$. We know that there are injections $p_m \to p_{m+1}$ and $p_{m+1} \to p_{m+2}$ and so on until $p_{n-2} \to p_{n-1}$, so composing them yields an injection $p_m \to p_{n-1}$. By Problem 2.45, there is a surjection $g : p_{n-1} \to p_m$. By composing $f \circ g$, we get a surjection $p_{n-1} \to p_n$, which contradicts Cantor's theorem. Thus, no lower set can be mapped to a higher set via a surjection, thus creating an infinite strict hierarchy of infinite sets. □

Corollary 2.55 (Cantor's paradox)**.** As a consequence of Cantor's theorem, we get a new proof that there is no set of all sets.

Proof. Suppose, for contradiction, that there exists a set V of all sets. Then the power set $\mathcal{P}(V)$ exists. The idea is to construct a surjection from V to $\mathcal{P}(V)$, which will contradict Cantor's theorem. Since every element of $\mathcal{P}(V)$ is a set, every element of $\mathcal{P}(V)$ is an element of V. So we can construct a surjection $f : V \to \mathcal{P}(V)$ by mapping the elements of $\mathcal{P}(V)$ (which are all elements of V) from V to $\mathcal{P}(V)$ by the identity function, and mapping any other elements of V to $\mathcal{P}(V)$ arbitrarily. For example, we can define

$$f(x) = \begin{cases} x & \text{if } x \in \mathcal{P}(V) \\ \varnothing & \text{if } x \notin \mathcal{P}(V) \end{cases}.$$

This is a surjection because each element of $\mathcal{P}(V)$ gets hit at least once. □

Problem 2.56. When I was a student in high school, I used to play with a Rubik's cube at school, while riding on public transit, and everywhere else. On several such occasions, people came up to me and mentioned that they had previously managed to "solve" five of the six faces (meaning for each of those five faces, its nine squares had stickers of the same colour), but that they had been unable to figure out how to solve the sixth face. There is no method of solving the Rubik's cube of which I am aware that proceeds by solving faces. This is because the cube is made up of pieces, including edges that have two stickers that are always adjacent to each other and corners that have three stickers that are always together. Methods for solving the cube are more about placing the edges and corners into the correct slots. Even without this knowledge, how did I conclude that these people were lying?

2.3 Induction

There exist several standard methods of proof, including direct proofs, proof by contrapositive, proof by exhaustion or casework, and proof by contradiction. There is one more powerful technique called induction that we can bring into play, thanks to our knowledge of sets.

Definition 2.57. For each positive integer n, the notation $[n]$ refers to the set

$$\{1, 2, \ldots, n\},$$

which is called a **section** of the positive integers $\mathbb{Z}_+ = \{1, 2, 3, \ldots\}$. Similarly, for each non-negative integer n, the notation $[n]^*$ refers to the section

$$\{0, 1, 2, \ldots, n\}$$

of the non-negative integers $\mathbb{Z}_{\geq 0} = \{0, 1, 2, 3, \ldots\}$.

Theorem 2.58. The following three assertions are all true and equivalent to each other.

1. **Well-ordering principle**: Let X be a non-empty subset of the integers $\mathbb{Z}$ such that X has a lower bound, meaning
$$\exists b \in \mathbb{Z}, \forall x \in X, x \geq b.$$
Then X has a minimal or least element, meaning
$$\exists m \in X, \forall x \in X, x \geq m.$$
Thanks to the antisymmetry of real inequalities, if there are two minimal elements m_1, m_2, they must be equal (due to $m_1 \geq m_2$ and $m_2 \geq m_1$), so the minimal element is unique.

2. **Principle of mathematical induction**: Let X be a subset of the positive integers $\mathbb{Z}_+$ such that $1 \in X$ and
$$\forall n \in \mathbb{Z}_+, (n \in X \implies n+1 \in X).$$
Then $X = \mathbb{Z}_+$. One can think of this as a domino effect.

3. **Principle of complete induction**: Let X be a subset of the positive integers $\mathbb{Z}_+$ such that $1 \in X$ and
$$\forall n \in \mathbb{Z}_+, ([n] \subseteq X \implies n+1 \in X).$$
Then $X = \mathbb{Z}_+$. This is a version of induction where we can assume more in the hypothesis of the second condition, which gives us more power.

The proof of this tri-fold equivalence is a bit too technical for our purposes. It is often given in introductory courses in theoretical computer science.

True understanding of the well-ordering principle and the two forms of induction listed will have to await the various examples that will occur in practice. The well-ordering principle is quite useful in number theory, as we will see in the proofs of Euclidean division, Bézout's lemma, and the fundamental theorem of arithmetic in Volume 3. In Volume 2, we use it to prove that the Calkin-Wilf tree bijectively enumerates the rationals.

Corollary 2.59. Let Y be a subset of $\mathbb{Z}$ such that Y has an upper bound c, meaning
$$\exists c \in \mathbb{Z}, \forall y \in Y, y \leq c.$$
Then Y has a maximal element n, meaning
$$\exists n \in Y, \forall y \in Y, y \leq n.$$

Proof. The idea is to use the well-ordering principle on another set where the upper bound is converted into a lower bound. Suppose Y is a subset of $\mathbb{Z}$ with upper bound c. Then $y \leq c$ for every $y \in Y$. The idea is to use a "reflection" to rewrite this as $1 \leq 1 + c - y$. Then the set
$$1 + c - Y = \{1 + c - y : y \in Y\}$$
is a subset of $\mathbb{Z}_+$ and so it has a lower bound of 1. By the well-ordering principle, $1 + c - Y$ has a least element x_0. So,
$$\forall y \in Y, 1 + c - y \geq x_0,$$
which we can rewrite as $\forall y \in Y, y \leq 1 + c - x_0$. So $1 + c - x_0$ is an upper bound for Y and it suffices to show that $1 + c - x_0 \in Y$. Indeed, since $x_0 \in 1 + c - Y$, there exists a $y_0 \in Y$ such that $x_0 = 1 + c - y_0$, which we can rewrite as $1 + c - x_0 = y_0 \in Y$, as desired. □

Problem 2.60. A "weaker" version of the well-ordering principle states that, if X is a non-empty subset of $\mathbb{Z}_+$ (emphasis on the positive sign), then X has a least element. Show that this version is not really weaker by proving that it implies the version stated in Theorem 2.58.

The well-ordering principle is a highly useful method by itself that does not usually need any modification. We will, however, combine the well-ordering principle with proof by contradiction when we study proof by infinite descent in Diophantine analysis in Volume 3.

Theorem 2.61 (Proof by induction)**.** Let $P(n)$ be a predicate in the variable n which ranges over the positive integers. Then:

1. If $P(1)$ is true and, for every positive integer n, the implication $P(n) \implies P(n+1)$ is true, then $P(n)$ is true for all positive integers n. This is called (ordinary) **induction**.

2. If $P(1)$ is true and, for every positive integer n, the implication

$$(P(1) \wedge P(2) \wedge \cdots \wedge P(n)) \implies P(n+1)$$

is true, then $P(n)$ is true for all positive integers. This is called **strong** or **complete induction**.

Here, $P(1)$ is called the **base case** and either of the two implications is called the **inductive step**, with the premise of the inductive step being called the **induction hypothesis**.

Proof. Using set builder notation, let

$$S = \{n \in \mathbb{Z}_+ : P(n) \text{ is true}\}.$$

1. Suppose $P(1)$ is true and that $P(n) \implies P(n+1)$ is true for every positive integer n. This means $1 \in S$ and

$$\forall n \in \mathbb{Z}_+, (n \in S \implies n+1 \in S).$$

By the principle of mathematical induction, $S = \mathbb{Z}_+$, so $\forall n \in \mathbb{Z}_+$, $P(n)$ is true.

2. Suppose $P(1)$ is true and that

$$(P(1) \wedge P(2) \wedge \cdots \wedge P(n)) \implies P(n+1)$$

is true for every positive integer n. This means $1 \in S$ and

$$\forall n \in \mathbb{Z}_+, ([n] \subseteq S \implies n+1 \in S).$$

By the principle of complete induction, $S = \mathbb{Z}_+$ again.

□

There is another form of induction, called structural induction, which applies to recursively defined sets, but we will not delve into it.

Example 2.62. For each positive integer n, it holds that

$$1 + 2 + \cdots + n = \frac{n(n+1)}{2}.$$

Solution. We will use induction. The base case is true because

$$1 = \frac{1(1+1)}{2}.$$

Suppose the sum holds for some positive integer n. So we are assuming that

$$1 + 2 + \cdots + n = \frac{n(n+1)}{2}.$$

Adding $n+1$ to both sides yields

$$1 + 2 + \cdots + n + (n+1) = \frac{n(n+1)}{2} + (n+1) = \frac{n(n+1) + 2(n+1)}{2} = \frac{(n+1)(n+2)}{2},$$

which finishes the inductive step and completes the proof by induction. This solution was rather mechanical. We will provide a more illuminating proof when we study arithmetic series in Theorem 7.3. □

Here are some tips with regards to proof by induction:

1. Mathematical induction is different from scientific induction. The former is a rigorous tool pertaining to the mathematical universe, whereas the latter is a tool for extrapolating evidence into general assumptions about the physical world.

2. It might be necessary to manually verify several base cases before an inductive step can be applied conveniently. That is, verify $P(1), P(2), \ldots, P(m)$ manually for some positive integer m, and then show that $P(n) \implies P(n+1)$ for all integers $n \geq m$. In fact a strong induction argument might require that the previous m cases are known to hold, such as in the theory of linear recursions, as studied in Volume 2.

3. Instead of $\mathbb{Z}_+$, it is fine to iterate over a different "countable" set, like the even integers, because the definition of countability of a non-empty set is that the set is in bijection with $\mathbb{Z}_+$.

4. Regardless of which variant of induction is used, what is of utmost importance is that the argument actually induces a domino effect. For example, it is advisable to manually verify that that induction step works when $P(1)$ is the induction hypothesis. That is, we need to be able to go from the base case to the first non-base case. This is the most precarious step and the source of fallacies.

5. On the whole, think of induction as allowing us to know the truth of some statements that should have taken an infinite lifetime to verify, even though we live finite lives. This is the power of mathematics. The idea is similar to the universal generalization rule of inference for quantifiers, where we fix an arbitrary general element and prove something about it, and all of a sudden the property holds for all such elements.

Problem 2.63. Prove that, for each positive integer n,

$$\frac{1}{1^2} + \frac{1}{2^2} + \frac{1}{3^2} + \cdots + \frac{1}{n^2} \leq 2 - \frac{1}{n}.$$

Since these partial sums are strictly increasing in value while being bounded above by 2, this proves that the infinite sum

$$\frac{1}{1^2} + \frac{1}{2^2} + \frac{1}{3^2} + \cdots$$

converges to some number less than or equal to 2.

Problem 2.64. Prove that, for every positive integer n, the following double inequality holds:

$$\sqrt{n} \leq \frac{1}{\sqrt{1}} + \frac{1}{\sqrt{2}} + \frac{1}{\sqrt{3}} + \cdots + \frac{1}{\sqrt{n}} \leq 2\sqrt{n}.$$

Problem 2.65. For positive integers n, find a closed formula for the sum

$$\frac{1}{2!} + \frac{2}{3!} + \frac{3}{4!} + \cdots + \frac{n}{(n+1)!}.$$

As a hint, evaluate the expression for $n = 1, 2, 3$ and find a pattern which can be proven by induction. Here, the factorial $n!$ refers to the product $1 \cdot 2 \cdots (n-1)n$.

Chapter 3

Arithmetic

"God made the integers, all else is the work of man."

– *Leopold Kronecker*

"We are justified in calling numbers a free creation of the human mind."

– *Richard Dedekind, Essays on the Theory of Numbers*

The study of algebra is preceded by the study of numbers and their associated operations, as in arithmetic. We will study abstract operations first, and then move into the standard addition, multiplication, subtraction, and division of real numbers. Afterwards, we will complete our study of arithmetic by looking at exponents, radicals, and logarithms.

3.1 Operations

We use binary operations on a daily basis, but without thinking of them as functions. Binary operations are about putting together two objects of the same type to produce a third object of the same type. We will start by studying operations abstractly as functions.

Definition 3.1. Given a set S, a **binary operation** on S is a function $f : S \times S \to S$. Instead of denoting it as $f(x, y)$, a symbol like $\star$ or $\circ$ (unrelated to function composition) denoting f is placed in between the two ordered inputs like $x \star y$. This is called **infix notation**, as opposed to the usual **prefix notation** of functions.

Example. Addition and multiplication are examples of binary operations on $\mathbb{R}$. Of course, we denote addition as $a + b$ instead of $+(a, b)$, and multiplication as $a \times b$ instead of $\times(a, b)$.

Theorem 3.2. If $a = b$ is an equation, then we may apply a function f that has a and b in its domain to get $f(a) = f(b)$. For example, this works with binary operations where one entry is fixed. That is, if $\star$ is a binary operation on a set S and $a, b, c \in S$, then

$$a = b \implies c \star a = c \star b,$$
$$a = b \implies a \star c = b \star c.$$

However, "doing the same thing to both sides" in a way that what we are applying is not a function does not necessarily work; for example, taking the numerators of the true equation $\frac{1}{2} = \frac{2}{4}$ yields the false equation $1 = 2$.

Proof. The result does not need much justification. Simply note that the same function being applied to equal inputs (so the inputs are the same objects) yields the same output for both by the functional property of a function (see Definition 2.24). In the special case of a fixed entry, we know that $\star(\cdot, \cdot)$ is a function from $S \times S$ to S, so fixing the first or second entry produces a function from S to S. □

Corollary 3.3. Let $\star$ be a binary operation on a set S and $a, b, c, d \in S$. If $a = b$ and $c = d$, then

$$a \star c = b \star d.$$

This is a very useful result in arithmetic because it allows us to "smash" equations together.

Proof. This is proven by the sequence

$$\begin{cases} a = b & \implies a \star c = b \star c \\ c = d & \implies b \star c = b \star d \end{cases} \implies a \star c = b \star d.$$

We have used transitivity of equality here. □

Binary operations might have one or more of the following convenient properties.

Definition 3.4. Let $\circ$ and $\star$ be binary operations on a set S.

- It is said that $\circ$ is **commutative** if, for all $a, b \in S$,

$$a \circ b = b \circ a.$$

- It is said that $\circ$ is **associative** if, for all $a, b, c \in S$,

$$a \circ (b \circ c) = (a \circ b) \circ c.$$

- It is said that $\star$ is **left-distributive** over $\circ$ if, for all $a, b, c \in S$,

$$c \star (a \circ b) = (c \star a) \circ (c \star b)$$

  and $\star$ is **right-distributive** over $\circ$ if, for all $a, b, c \in S$,

$$(a \circ b) \star c = (a \star c) \circ (b \star c).$$

  If $\star$ is both left-distributive and right-distributive over $\circ$, then we say that $\star$ is **distributive** over $\circ$.

Example. Addition and multiplication of real numbers are both commutative and associative, and multiplication is distributive over addition. Subtraction and division are neither commutative nor associative. Those who are familiar with matrix multiplication will recognize that it is associative but not commutative (Definition A.2). An example of a commutative but not associative operation is the midpoint operation $(a, b) \mapsto \dfrac{a+b}{2}$ on real numbers.

Problem 3.5. Write the definition of commutativity, associativity, and each kind of distributivity in prefix function notation, like $\circ(a, b)$. This exercise should give the reader an appreciation for why the usual notation is much easier for the human mind to grasp in practice.

Theorem 3.6. Let $\star$ and $\circ$ be binary operations on S. If $\star$ is commutative, then $\star$ is left-distributive over $\circ$ if and only if $\star$ is right-distributive over $\circ$.

Proof. Suppose $\star$ is commutative. Then $\star$ is left-distributive over $\circ$ if and only if for all $a, b, c \in S$,

$$c \star (a \circ b) = (c \star a) \circ (c \star b).$$

By applying commutativity of $\star$ to each of the three applications of $\star$, this is equivalent to

$$(a \circ b) \star c = (a \star c) \circ (b \star c),$$

which is the definition of right-distributivity. We can go in the other direction as well. □

Definition 3.7. Let $\circ$ be a binary operation on S. An element $e \in S$ is an **identity** for $\circ$ if for all $a \in S$,

$$e \circ a = a \circ e = a.$$

Example. For the addition of real numbers, 0 is an identity. For the multiplication of real numbers, 1 is an identity. An arbitrary binary operation does not necessarily have an identity.

Theorem 3.8. Let $\circ$ be a binary operation on S. If $\circ$ has an identity, then the identity is unique.

Proof. Suppose $\circ$ has two identities $e_1, e_2 \in S$. Since e_1 is an identity and e_2 is an element of S, $e_1 \circ e_2 = e_2$. Since e_2 is an identity and e_1 is an element of S, $e_1 \circ e_2 = e_1$. Thus,

$$e_1 = e_1 \circ e_2 = e_2,$$

meaning e_1 and e_2 are the same element. □

Definition 3.9. Let $\circ$ be a binary operation on S with an identity e. Then b is said to be a **left-inverse** of a if $b \circ a = e$, and b is said to be a **right-inverse** of a if $a \circ b = e$. If b is both a left-inverse and a right-inverse of a, then b is said to be an **inverse** of a.

Example. For the addition of real numbers, $-a$ is an inverse of a. For the multiplication of non-zero real numbers, $a^{-1} = \frac{1}{a}$ is a multiplicative inverse of a. For an arbitrary binary operation that has an identity, an arbitrary element of the underlying set does not necessarily have an inverse.

Theorem 3.10. Let $\circ$ be an associative binary operation on S with an identity. If an element a has an inverse, then the inverse of a is unique.

Proof. Let the identity of $\circ$ be e. Suppose a has two inverses b and c. Then

$$b = b \circ e = b \circ (a \circ c) = (b \circ a) \circ c = e \circ c = c,$$

so b and c are the same element. This proof is easier to comprehend if one begins with the central equality and then moves outward on the left and right separately. □

Definition 3.11. If $\star$ is a binary operation on a set S and R is a subset of S, then R is said to be **closed** under $\star$ if, for all $a, b, \in R$, it holds that $a \star b \in R$.

There are three abstract algebraic structures that will occasionally crop up in our studies. We will use groups to count under symmetry in Volume 2. Rings and fields will come up when studying polynomials in Chapter 11. A full course in university-level abstract algebra would be necessary to truly appreciate the definitions.

Definition 3.12. A **group** is an ordered pair $(G, \star)$ of a set G and a binary operation $\star$ on G such that $\star$ is associative, there is an identity element, and every element has an inverse. If $\star$ is also commutative, then the group is called **abelian**. Often, G itself is referred to as the group even though it is just the set component of the ordered pair that is the group.

Definition 3.13. A **ring** is an ordered triple $(R, \circ, \star)$ of a set R and binary operations $\circ$ and $\star$ such that $(R, \circ)$ is an abelian group, $\star$ is associative, and $\star$ is distributive over $\circ$. If $\star$ is also commutative, then the ring is called **commutative**. It is not necessary for $\star$ to have an identity the way $\circ$ does, but if $\star$ is known to have an identity, then it is good to mention it.

Definition 3.14. A **field** is a commutative ring such that the second operation has an identity and every element that is not the identity of the first operation has an inverse under the second operation. Alternatively formulated, a field is an ordered triple $(F, \circ, \star)$ of a set F and binary operations $\circ$ and $\star$ such that $(F, \circ)$ is an abelian group, and if x is the identity for $\circ$, then $(F \backslash \{x\}, \star)$ is also an abelian group, and $\star$ is distributive over $\circ$.

With the abstractions out of the way, let us focus on real numbers and their special subsets, the integers and the rationals.

Definition 3.15. We will not formally construct the **real numbers** out of sets, but they can be boiled down to sets. The reader should already be familiar with how real numbers behave since they fulfill our natural sense of quantification. There is one semi-mysterious property of the real numbers called Dedekind completeness (see Definition 5.19), but it does not matter much until one begins to learn calculus. The basic operations of real number arithmetic are **addition** and **multiplication**, which then lead to **subtraction** and **division**. The set of real numbers is denoted by $\mathbb{R}$, positive real numbers by $\mathbb{R}_+$, and non-negative real numbers by $\mathbb{R}_{\geq 0}$.

Students ordinarily take an analysis course at some point where they are told that real numbers can be defined as equivalence classes of Cauchy sequences of rationals. As this definition is beyond the scope of our writing, we will instead focus on learning about the rules that real numbers satisfy.

Theorem 3.16. For real numbers a, b, c:

- $a = b$ implies $a + c = b + c$ and $c + a = c + b$
- $a = b$ implies $ac = bc$ and $ca = cb$

The two rules above allow us to deduce two more rules for real numbers a, b, c, d:

- $a = b$ and $c = d$ together imply $a + c = b + d$
- $a = b$ and $c = d$ together imply $ac = bd$

Four rules analogous to the above four rules hold for subtraction and for division by non-zero numbers, which will become evident when we define subtraction in terms of addition (Definition 3.22) and division in terms of multiplication (Definition 3.37).

Proof. This is a direct application of Theorem 3.2. □

Definition 3.17. Since we have numerous operations, we have to set down rules that govern how we interpret expressions that include operations. The mnemonic to remember for the order in which operations are interpreted is **PEMDAS**, which stands for: Parentheses, Exponents, Multiplication, Division, Addition, Subtraction. Some people use **BEDMAS** where the B stands for Brackets.

1. Firstly, if there are parentheses, we evaluate what is inside them first. If there are nested parentheses, meaning sets of parentheses within sets of parentheses, we evaluate them from the innermost parentheses outward. For example, in $(1 + 2) \cdot 3$, the $1 + 2$ is evaluated first.
2. We will describe exponents and how to work with them in Theorem 3.50.
3. Then come multiplication and division. For example, in $1 + 2 \cdot 3$, the $2 \cdot 3$ is evaluated first. We will see that division boils down to multiplication, so they are of equal rank.
4. Of lowest rank are addition and subtraction. We will see that subtraction boils down to addition, so they are of equal rank.

Example. There are some "puzzles" that have garnered attention in recent times such as the following: compute $3 \div 3 \cdot 3$. There are two possible interpretations:

$$(3 \div 3) \cdot 3 = 1 \cdot 3 = 3,$$
$$3 \div (3 \cdot 3) = 3 \div 9 = \frac{1}{3}.$$

An issue with such problems is the ambiguity stemming from the usage of the division symbol instead of expressing the division as a fraction. Secondly, in PEMDAS and BEDMAS, it is not clarified whether multiplication or division has priority, nor is it necessary. Simply use parentheses to avoid confusion if needed. The burden of clarity is on the writer.

Theorem 3.18. Let a, b, c be any real numbers. Then the following rules hold:

1. Addition is commutative: $a + b = b + a$
2. Addition is associative: $a + (b + c) = (a + b) + c$
3. Multiplication is commutative: $ab = ba$
4. Multiplication is associative: $a(bc) = (ab)c$
5. Zero is an additive identity: $a + 0 = 0 + a = a$

6. One is a multiplicative identity: $1 \cdot a = a \cdot 1 = a$
7. Multiplication distributes over addition: $a(b+c) = ab + ac$

Depending on the context, the word **trivial** might refer to the 0 or 1 if we are talking about a trivial number.

We will take these properties for granted, as a subset of the axioms for the real numbers. Most of the properties should be intuitively clear by interpreting $a+b$ as starting from a on the number line and moving away by b units, and $a \cdot b$ as meaning "a copies of b" in the case of positive integers.

Definition 3.19. When we use the distributive property to go from $a(b+c)$ to $ab+ac$, it is called **expanding**. The reverse process of going from $ab+ac$ to $a(b+c)$ is called **factoring**.

Theorem 3.20. *For every real number a, there is a unique real number $-a$, such that $a + (-a) = 0$. We call $-a$ the **additive inverse** or **negative** of a.*

It should be intuitively clear that $-a$ is the mirror image of a across 0 on the number line.

Definition 3.21. The essence of the **integers** is that repeatedly adding 1 (which represents singleness) to itself produces all of the positive integers, that there is a 0 integer that represents "nothing," and the negative integers are the additive inverses of the positive integers. The set of integers is denoted by $\mathbb{Z}$, the **positive integers** by $\mathbb{Z}_+$, and the set of **non-negative integers** by $\mathbb{Z}_{\geq 0}$, where non-negative means the positives along with zero. The integers form a group under addition, and a commutative ring under addition and multiplication together.

Definition 3.22. We define **subtraction** by b as the addition of the negative of b. In other words, $a-b$ is defined as $a+(-b)$. Note that subtraction is not associative because

$$(a-b)-c \neq a-(b-c)$$

unless $c = 0$. Subtraction is also not commutative. Moreover, to be clear, we interpret the expression $a-b-c$ as applying operations from left to right, so that it is $a+(-b)+(-c)$.

Lemma 3.23 (Additive cancellation law)**.** *If a, b, c are real numbers such that*

$$a + c = b + c,$$

then $a = b$.

Proof. We can add $-c = -c$ to $a+c = b+c$ to get $a = b$. This is allowed because we established in Theorem 3.16 that equations may be added to each other. □

Theorem 3.24. *For any real number a, it holds that $a \cdot 0 = 0$.*

Proof. Since 0 is the additive identity, $0 + 0 = 0$. Multiplying both sides by a and using the distributive law yields

$$a \cdot 0 = a \cdot (0+0) = a \cdot 0 + a \cdot 0.$$

Now we can subtract $a \cdot 0 = a \cdot 0$ from both sides to get $a \cdot 0 = 0$, as desired. □

Theorem 3.25. *For any real number a, it holds that $-a = (-1) \cdot a$. So, taking the negative of a number is equivalent to multiplying it by the negative of the multiplicative identity.*

Proof. By Theorem 3.20, $-a$ is the unique number that satisfies the equation $a + (-a) = 0$. So -1 is the number that satisfies the equation $1 + (-1) = 0$. Multiplying both sides by a and using the distributive law yields

$$a + (-1) \cdot a = 0.$$

Equating this with $a + (-a) = 0$ yields

$$a + (-a) = a + (-1) \cdot a$$

and subtracting a from both sides gives $-a = (-1) \cdot a$. □

Corollary 3.26. For any real numbers a and b, it holds that $(-a)b = -(ab)$. Consequently, $a(-b)$ represents the same number. So, the expression $-ab$ is well-defined, as it can refer to any of $(-a)b$ or $-(ab)$ or $a(-b)$.

Proof. Since $-c = (-1) \cdot c$ holds for any real number c, we find that, by associativity,

$$(-a)b = ((-1)a)b = (-1)(ab) = -(ab).$$

For the consequence,

$$(-a)b = -(ab) = -(ba) = (-b)a = a(-b).$$

□

Corollary 3.27. For any real numbers a and b, it holds that $(-a)(-b) = ab$. As the well-known saying goes, "double negatives makes a positive."

Proof. Since $a+(-a) = 0$ and $b+(-b) = 0$, multiplying the equations and using the distributive law yields:

$$\begin{aligned} (a + (-a))(b + (-b)) &= 0 \cdot 0 \\ ab + a(-b) + (-a)b + (-a)(-b) &= 0 \\ ab + (-ab) + (-ab) + (-a)(-b) &= 0 \\ -ab + (-a)(-b) &= 0 \\ (-a)(-b) &= ab. \end{aligned}$$

In the last step, we added ab to both sides. □

Problem 3.28. For a real number a, what is the negative of the negative of a? Justify it.

Definition 3.29. The **absolute value** of a real number is its distance from 0 on the real number line. More precisely, we define it as

$$|x| = \begin{cases} x & \text{if } x \text{ is non-negative} \\ -x & \text{if } x \text{ is negative} \end{cases}.$$

Note that, if x is negative, then tacking on an extra negative sign as shown makes the output positive, by the solution to Problem 3.28. Absolute value is a very useful function that we will study in detail in Section 6.1.

Definition 3.30. There are two basic kinds of symmetry for functions f whose domain and range are subsets of $\mathbb{R}$ such that $x \in \text{Dom}(f) \iff -x \in \text{Dom}(f)$:

- The function f is **even** if $f(-x) = f(x)$ for all $x \in \text{Dom}(f)$.
- The function f is **odd** if $f(-x) = -f(x)$ for all $x \in \text{Dom}(f)$.

It becomes evident why we refer to these definitions as instances of symmetry if one graphs even or odd functions on the Cartesian plane.

Example. Even and odd functions are defined as such because the function $f(x) = x^n$ with domain $\mathbb{R}$ is even for even integers n and odd for odd integers n. We will explain exponents in Section 3.2, for those looking for a definition. The graphs of x^2 and x^3 are given below on the left and right, respectively.

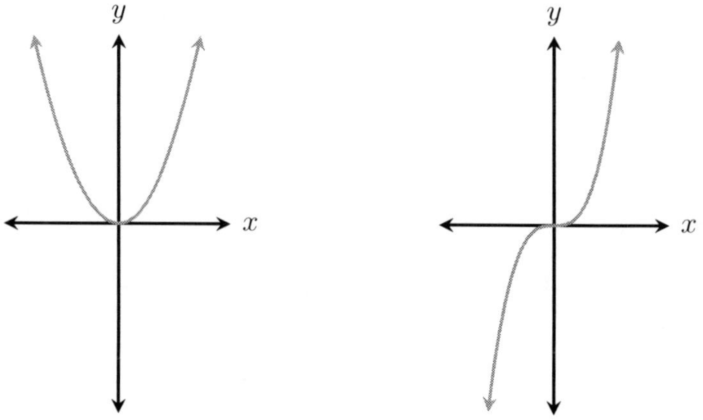

Even and odd functions have interesting properties with regards to their sums, differences, products, quotients and compositions. These properties are too numerous to exhaustively list here, but the reader is encouraged to look into them elsewhere. We have mentioned a couple of the more interesting properties in Example 3.31 and Problem 3.32.

Example 3.31. Let f be a function whose domain and range are subsets of $\mathbb{R}$ such that $x \in \mathrm{Dom}(f) \iff -x \in \mathrm{Dom}(f)$. Show that there exists a unique even function f_e and a unique odd function f_o, both with domains equal to $\mathrm{Dom}(f)$, such that

$$f(x) = f_e(x) + f_o(x)$$

for all $x \in \mathrm{Dom}(f)$.

Solution. We will first prove uniqueness of f_e and f_o. If f_e and f_o exist, then

$$\begin{aligned} f(x) &= f_e(x) + f_o(x) \\ f(-x) &= f_e(-x) + f_o(-x) \\ &= f_e(x) - f_o(x). \end{aligned}$$

Solving this system for f_e and f_o yields

$$\begin{aligned} f_e(x) &= \frac{f(x) + f(-x)}{2}, \\ f_o(x) &= \frac{f(x) - f(-x)}{2} \end{aligned}$$

for all $x \in \mathrm{Dom}(f)$. This proves uniqueness, since f_e and f_o must be two specific functions in terms of f. Moreover, it can be readily verified that f_e is even and f_o is odd, and that $f = f_e + f_o$, which proves existence. □

Problem 3.32. Find all functions $f : \mathbb{R} \to \mathbb{R}$ that are both even and odd.

Theorem 3.33. For every non-zero real number a, there is a unique real number a^{-1}, such that $a \cdot a^{-1} = 1$. We call a^{-1} the **multiplicative inverse** or **reciprocal** of a.

The intuitive idea behind reciprocals is that a segments of length a^{-1} on the number line come together to make a segment of length 1. In other words, if a segment of 1 were to be divided into a segments, then each small segment would have length a^{-1}.

Definition 3.34. The **rationals** can be produced by taking the multiplicative inverse of each non-zero integer, and then adding each such inverse to itself arbitrarily many finite times. Since repeated addition is simply multiplication, set builder notation gives us

$$\mathbb{Q} = \{ab^{-1} : a \in \mathbb{Z}_{\geq 0}, b \in \mathbb{Z} \backslash \{0\}\}.$$

The rationals are denoted by $\mathbb{Q}$, the **positive rationals** by $\mathbb{Q}_+$, and the **non-negative rationals** by $\mathbb{Q}_{\geq 0}$. The rationals form a field, just like the reals.

When we get to complex numbers, we will see that they form a field too (Theorem 9.5).

Lemma 3.35 (Multiplicative cancellation law)**.** If a, b, c are real numbers such that $ac = bc$ and c is non-zero, then $a = b$.

Proof. We can multiply $ac = bc$ by $c^{-1} = c^{-1}$ to get $a = b$. This is allowed because we established in Theorem 3.16 that equations may be multiplied together. □

Theorem 3.36. If a and b are real numbers such that $ab = 0$, then $a = 0$ or $b = 0$. For this reason, we often like to take every term in an equation to one side and factor it so that we have a factored expression equal to 0. This allows us to deduce that at least one of the factors is equal to 0.

Proof. Assuming $ab = 0$, it suffices to show that $b \neq 0$ implies $a = 0$. Since b is non-zero, b has a multiplicative inverse b^{-1}. By multiplying $ab = 0$ by $b^{-1} = b^{-1}$, we get

$$(ab)b^{-1} = 0 \cdot b^{-1} \implies a = 0.$$

So $b = 0$ or $a = 0$. Here, we have used Theorem 3.24. □

Definition 3.37. We define **division** by a non-zero real number b as the multiplication by multiplicative inverse of b. In other words, $\frac{a}{b}$ is defined as $a \cdot b^{-1}$. Note that this means $a^{-1} = 1 \cdot a^{-1} = \frac{1}{a}$, which is called the **reciprocal** of a. The division expression $\frac{a}{b}$ is called a **fraction**. The fraction bar, called a **vinculum**, that separates the **numerator** a and **denominator** b implicitly says that we are putting parentheses around the numerator and around the denominator. For example, $\frac{a+b}{c+d}$ means $\frac{(a+b)}{(c+d)}$. Unlike subtraction, an expression such as $a/b/c$ without parentheses has no commonly accepted definition and so it is ill-defined; it could represent $\frac{a}{\left(\frac{b}{c}\right)}$ or $\frac{\left(\frac{a}{b}\right)}{c}$.

Problem 3.38. Compute $\frac{-1}{-1}$.

Lemma 3.39. For any non-zero real numbers a and b, it holds that $(ab)^{-1} = a^{-1}b^{-1}$.

Proof. Indeed, we find that

$$(ab)(a^{-1}b^{-1}) = (ab)(b^{-1}a^{-1}) = a(bb^{-1})a^{-1} = aa^{-1} = 1$$

so the multiplicative inverse of ab is $a^{-1}b^{-1}$. This is reminiscent of Theorem 2.33, though we had the added power of commutativity just now. □

Problem 3.40. Similar to taking the negative of a negative (see Problem 3.28), show that applying the multiplicative inverse twice to a non-zero real number returns the original number. This means $(a^{-1})^{-1} = a$ for $a \neq 0$.

Theorem 3.41. Fractions have several properties that are useful in computations and simplifications. For real numbers a, b, c, d such that c and d are non-zero, the following identities hold with any additional restrictions stated:

1. Addition and subtraction: $\frac{a}{c} \pm \frac{b}{c} = \frac{a \pm b}{c}$, where the two $\pm$ signs match
2. Multiplication: $\frac{a}{c} \cdot \frac{b}{d} = \frac{ab}{cd}$
3. Simplification: $\frac{ad}{cd} = \frac{a}{c}$
4. Reciprocal: $\left(\frac{a}{c}\right)^{-1} = \frac{c}{a}$ if $a \neq 0$
5. Division: $\frac{\left(\frac{a}{c}\right)}{\left(\frac{b}{d}\right)} = \frac{ad}{bc}$ if $b \neq 0$. The following special cases are useful in our experience:

$$\frac{\left(\frac{a}{b}\right)}{c} = \frac{a}{bc}, \quad \frac{a}{\left(\frac{b}{c}\right)} = \frac{ac}{b},$$
$$\frac{\left(\frac{1}{a}\right)}{b} = \frac{1}{ab}, \quad \frac{a}{\left(\frac{1}{b}\right)} = ab,$$
$$\frac{1}{\left(\frac{a}{b}\right)} = \frac{b}{a}, \quad \frac{1}{\left(\frac{1}{a}\right)} = a,$$
$$\frac{\left(\frac{a}{c}\right)}{\left(\frac{b}{c}\right)} = \frac{a}{b}, \quad \frac{\left(\frac{c}{a}\right)}{\left(\frac{c}{b}\right)} = \frac{b}{a}.$$

In general, it is not true that $\dfrac{a+b}{c+d} = \dfrac{a}{c} + \dfrac{b}{d}$.

Proof. The identities can be deduced from the definitions of inverses and fractions, along with Lemma 3.39, Problem 3.40, and the distributive property. At times, we will also use the earlier identities in this list to prove later identities in this list.

1. $\dfrac{a}{c} \pm \dfrac{b}{c} = ac^{-1} \pm bc^{-1} = (a \pm b)c^{-1} = \dfrac{a \pm b}{c}$

2. $\dfrac{a}{c} \cdot \dfrac{b}{d} = (ac^{-1})(bd^{-1}) = (ab)(c^{-1}d^{-1}) = (ab)(cd)^{-1} = \dfrac{ab}{cd}$

3. $\dfrac{ad}{cd} = (ad) \cdot (cd)^{-1} = adc^{-1}d^{-1} = ac^{-1}dd^{-1} = ac^{-1} = \dfrac{a}{c}$

4. $\dfrac{a}{c} \cdot \dfrac{c}{a} = \dfrac{ac}{ca} = \dfrac{a}{a} = 1$, so $\dfrac{c}{a}$ is the multiplicative of inverse of $\dfrac{a}{c}$

5. $\dfrac{\left(\frac{a}{c}\right)}{\left(\frac{b}{d}\right)} = \dfrac{a}{c} \cdot \left(\dfrac{b}{d}\right)^{-1} = \dfrac{a}{c} \cdot \dfrac{d}{b} = \dfrac{ad}{bc}$

By choosing $a = b = c = d = 1$, we find that $\dfrac{1+1}{1+1} = 1 \neq 2 = \dfrac{1}{1} + \dfrac{1}{1}$. □

Definition 3.42. The act of turning the sum $\dfrac{a}{b} + \dfrac{c}{d}$ into the single equivalent fraction $\dfrac{ad+bc}{bd}$ is called **finding a common denominator** and bd is called the **common denominator** in this case. Any other denominator that allows for turning the sum of multiple fractions into one fraction also fulfills this definition.

Definition 3.43. Given several fractions with integer denominators, the **lowest common denominator** is the least common multiple of the integer denominators, and the number may be used as a common denominator to add the fractions together.

Problem 3.44. Prove the following identities for real numbers a and $b \neq 0$:

- $\dfrac{-a}{-b} = \dfrac{a}{b}$
- $\dfrac{-a}{b} = -\left(\dfrac{a}{b}\right) = \dfrac{a}{-b}$

As such, in addition to Definition 3.34, we may equivalently define the rationals as

$$\mathbb{Q} = \left\{\frac{a}{b} : a \in \mathbb{Z}, b \in \mathbb{Z}\backslash\{0\}\right\}.$$

Definition 3.45. If there are one or more fractions in an equation, then the act of multiplying both sides of the equation by the product of the denominators, in order to produce an equation with no fractions, is called **clearing the denominators**. If the denominators are all integers, we can get away with multiplying by only their least common multiple.

Theorem 3.46. If a, b, c, d are real numbers such that $\dfrac{a}{b} = \dfrac{c}{d}$ (so b and d must be non-zero), then

$$\frac{a}{b} = \frac{c}{d} = \frac{a+c}{b+d},$$

assuming $b \neq -d$. Similarly, if $b \neq d$, then $\dfrac{a}{b} = \dfrac{c}{d} = \dfrac{a-c}{b-d}$.

Proof. We will work backwards using reversible steps:

$$\begin{aligned} \frac{a}{b} &= \frac{a+c}{b+d} \\ a(b+d) &= b(a+c) \\ ab+ad &= ba+bc \\ ad &= bc \\ \frac{a}{b} &= \frac{c}{d}, \end{aligned}$$

which we have assumed to be true. For the second result, we can apply the first result as follows:

$$\frac{a}{b} = \frac{c}{d} = \frac{-c}{-d} \implies \frac{a-c}{b-d} = \frac{a+(-c)}{b+(-d)} = \frac{a}{b} = \frac{-c}{-d} = \frac{c}{d}.$$

□

The reader is expected to be familiar with how to apply arithmetic operations to integers and rationals using standard computational algorithms.

Problem 3.47. Note that $\frac{a}{b} = \frac{c}{d}$ if and only if $ad = bc$, assuming the denominators are never 0. This leads us to defining the **ratio binary relation** on $(\mathbb{R} \times \mathbb{R}) \backslash \{(0,0)\}$, which is defined by

$$(a,b) \sim (c,d) \iff ad = bc.$$

Prove that this is an equivalence relation. The equivalence class of (a, b) is represented by the **ratio** $a : b$.

Problem 3.48. Reviewing Definition 2.36 if needed, show that the following functions are involutions.

1. $f : \mathbb{R} \to \mathbb{R}$ defined by $c - x$ where c is any real constant
2. $g : \mathbb{R}\backslash\{0\} \to \mathbb{R}\backslash\{0\}$ defined by $\frac{c}{x}$ where c is any non-zero real constant

3.2 Exponents, Radicals, Logarithms

Definition 3.49. The **exponential function** $f_b : D \to \mathbb{R}$ with a fixed real base b is a function that is denoted with input x by $f_b(x) = b^x$. For us, the domain is defined as

$$D = \begin{cases} \mathbb{R} & \text{if } b > 0 \\ \mathbb{R}_+ & \text{if } b = 0 \\ \mathbb{Z} \cup \left\{ \frac{m}{n} : m, n \in \mathbb{Z}, n \text{ is odd}, \gcd(m,n) = 1 \right\} & \text{if } b < 0 \end{cases}.$$

In the expression b^x, the whole expression b^x is called a **power**, b is called the **base**, and x is called the **exponent**. We will not formally define the exponential function, but we have noted numerous properties of powers in Theorem 3.50 that are convenient for computation. Going back to PEMDAS or BEDMAS, the E stands for "exponent" (Definition 3.17). This means that exponents have a higher rank than the four arithmetic operations.

Example. In $a^b + c$, the a^b is computed first and then c is added in. Another, more important, example is that $-a^n$ is not necessarily equal to $(-a)^n$, but rather it is the negative of a^n. This is because $-a^n$ means $(-1) \cdot a^n$, and multiplication comes after exponentiation.

Theorem 3.50. The exponential function and powers satisfy the following arithmetic properties:

1. The exponential function is bijective onto $\mathbb{R}_+$ when b is positive but not equal to 1. The exponential function is identically 1 for $b = 1$, and identically 0 for $b = 0$.

2. If n is a positive integer and b is real, then

$$b^n = \underbrace{b \cdot b \cdots b}_{n \text{ factors of } b}.$$

3. If $b \neq 0$ is real, then a power with its exponent equal to 0 can be computed as $b^0 = 1$. If $b = 0$, then $b^0 = 0^0$ is undefined due to the following dichotomy:

$$x \neq 0 \implies 0^x = 0,$$
$$b \neq 0 \implies b^0 = 1.$$

So it is not clear what 0^0 "should" be. Depending on the context, we might choose a temporary convention as appropriate and convenient. For example, we choose $0^0 = 1$ in the proof of Binet's formula for Fibonacci numbers and, more generally, when studying linear recurrences in Volume 2. This convention is also chosen for the Girard-Newton sums (Theorem 11.53) and several multivariable inequalities in Chapter 12.

4. If b is a positive real, then powers with negative exponents are defined in terms of powers with positive exponents as

$$b^{-x} = (b^x)^{-1} = \frac{1}{b^x}.$$

This is well-defined for positive b because $b^x = 0$ if and only if $b = 0$.

5. Powers with the same positive base b can be multiplied as

$$b^x \cdot b^y = b^{x+y}.$$

As a consequence, if $b \neq 0$, then we can divide powers with the same base as

$$\frac{b^x}{b^y} = b^x \cdot (b^y)^{-1} = b^x \cdot b^{-y} = b^{x-y}.$$

6. If a, b are positive reals, then powers with the same exponent can be multiplied as

$$a^x \cdot b^x = (ab)^x.$$

As a consequence, if $b \neq 0$, then we can divide powers with the same exponent as

$$\frac{a^x}{b^x} = a^x \cdot (b^x)^{-1} = a^x \cdot b^{-x} = a^x \cdot (b^{-1})^x = (ab^{-1})^x = \left(\frac{a}{b}\right)^x.$$

7. The power of a power with a positive base b can be computed as

$$(b^x)^y = (b^y)^x = b^{xy}.$$

This includes the rational exponent case

$$b^{\frac{p}{q}} = b^{p \cdot \frac{1}{q}} = (b^{\frac{1}{q}})^p = (b^p)^{\frac{1}{q}}$$

for integers p and $q \neq 0$. We show how to interpret and compute powers with rational exponents using radicals in Theorem 3.55. Note that there is a difference between $b^{(x^y)}$ and $(b^x)^y$. If no parentheses are used, then the convention is to interpret b^{x^y} as $b^{(x^y)}$.

If the base b is negative, there are analogues of most of these rules that still work, as long as the exponents are from the stated domain.

Example 3.51. Find a pattern in the sequence of powers of -1:

$$(-1)^1, (-1)^2, (-1)^3, (-1)^4, \ldots.$$

Solution. We leave it to the reader to prove by induction on $n \geq 1$ that

$$(-1)^n = \begin{cases} 1 & \text{if } n \text{ is even} \\ -1 & \text{if } n \text{ is odd} \end{cases}.$$

This pattern of alternating between 1 and -1 is frequently used as a tool in summation notation (Definition 4.10). By multiplying the term of index k by $(-1)^k$, we can indicate that the terms have alternating signs. For example,

$$\sum_{k=1}^{n} \frac{(-1)^k}{k} = -\frac{1}{1} + \frac{1}{2} - \frac{1}{3} + \cdots + \frac{(-1)^n}{n}.$$

□

Example 3.52. Let n be an odd positive integer. Show that the functions $f(x) = x^n$ and $g(x) = x^{\frac{1}{n}}$ are inverse functions of each other. To be clear, these are well-defined functions whose domains are $\mathbb{R}$.

Solution. Letting $m = \dfrac{1}{n}$, it holds that $mn = 1$. So we find that

$$f(g(x)) = (x^m)^n = x^{mn} = x^1 = x,$$
$$g(f(x)) = (x^n)^m = x^{nm} = x^1 = x.$$

Since both compositions yield the identity function, f and g are inverse functions. □

Theorem 3.53. Two tricks of the trade that are worth knowing are:

1. Adding b^x to itself n times, where n is a positive integer, yields nb^x. If $n = b$, then we get

$$\underbrace{b^x + b^x + \cdots + b^x}_{n \text{ copies of } b^x} = b \cdot b^x = b^{x+1}.$$

2. By distributivity,

$$b^x \pm b^y = b^y(b^{x-y} \pm 1).$$

This is usually useful for only $x \geq y$.

Definition 3.54. If n is a positive integer, then an nth **root** of a real number b is defined as any real (or complex) number r such that $r^n = b$. Complex numbers provides a complete theory of roots (see Theorem 9.30). Real numbers are messier, but it is thankfully true that every positive real b has a unique *positive* n^{th} root; we denote this root in **radical** notation as $\sqrt[n]{b}$ or, for $n = 2$, just $\sqrt{b}$. There are two more kinds of roots:

- For a positive real number b, the negative of $\sqrt[n]{b}$ satisfies

$$(-\sqrt[n]{b})^n = (-1)^n b.$$

This equals b for even positive integers n. Thus, $-\sqrt[n]{b}$ is called the **negative** n^{th} **root** of b if b is a positive real number and n is an even positive integer. The uniqueness of the positive root implies the uniqueness of the negative root (since they are negatives of each other), assuming the latter exists.

- For a negative real number b and odd positive integer n, there exists a unique real number r such that $r^n = b$. For example, $(-2)^3 = -8$, so we can write $\sqrt[3]{-8} = -2$.

Theorem 3.55. For a positive real b and a positive integer n,

$$\sqrt[n]{b} = b^{\frac{1}{n}}.$$

If we add a positive integer m exponent to the mix, we get that

$$\sqrt[n]{b^m} = b^{\frac{m}{n}}.$$

We can also bring a negative into the exponent by applying a reciprocal, like

$$\frac{1}{\sqrt[n]{b^m}} = \frac{1}{b^{\frac{m}{n}}} = b^{-\frac{m}{n}}.$$

Proof. Suppose b is a positive real number and n is a positive integer. Let $r = \sqrt[n]{b}$. Then

$$b = r^n \implies b^{\frac{1}{n}} = (r^n)^{\frac{1}{n}} = r^{n \cdot \frac{1}{n}} = r^1 = r = \sqrt[n]{b}.$$

For the second result, we can apply the first result and an exponent rule to get

$$\sqrt[n]{b^m} = (b^m)^{\frac{1}{n}} = b^{\frac{m}{n}}.$$

The third rule is evident from its statement. □

Theorem 3.56. If a and b are positive real numbers and n is a positive integer, then

$$\sqrt[n]{a} \cdot \sqrt[n]{b} = \sqrt[n]{ab},$$
$$\frac{\sqrt[n]{a}}{\sqrt[n]{b}} = \sqrt[n]{\frac{a}{b}} \text{ for } b > 0.$$

As a corollary, simplifications such as

$$\sqrt[n]{a^n b} = a\sqrt[n]{b}$$

are possible. Analogous rules hold when one or both of a, b are negative, as long as n is an odd positive integer.

Proof. Given positive real a, b and a positive integer n,

$$\sqrt[n]{a} \cdot \sqrt[n]{b} = a^{\frac{1}{n}} \cdot b^{\frac{1}{n}} = (ab)^{\frac{1}{n}} = \sqrt[n]{ab},$$
$$\frac{\sqrt[n]{a}}{\sqrt[n]{b}} = \frac{a^{\frac{1}{n}}}{b^{\frac{1}{n}}} = \left(\frac{a}{b}\right)^{\frac{1}{n}} = \sqrt[n]{\frac{a}{b}} \text{ for } b > 0.$$

We can apply the first rule to get that $\sqrt[n]{a^n b} = \sqrt[n]{a^n} \cdot \sqrt[n]{b} = a\sqrt[n]{b}$. In practice, extracting the largest n^{th} power a^n from under an nth root involves finding the prime factorization of the radical's argument. □

In general, when applying exponent rules, ensure that all expressions are defined. For example, there is no issue if a negative base is taken to a rational exponent whose denominator is odd, such as

$$(-8)^{\frac{2}{3}} = ((-8)^2)^{\frac{1}{3}} = 64^{\frac{1}{3}} = 4, \text{ or}$$
$$(-8)^{\frac{2}{3}} = ((-8)^{\frac{1}{3}})^2 = (-2)^2 = 4.$$

However, if the denominator is even, issues like

$$1 = 1^{\frac{1}{2}} = ((-1)(-1))^{\frac{1}{2}} = (-1)^{\frac{1}{2}}(-1)^{\frac{1}{2}}$$

can occur, where the right side is not even defined without complex numbers. Even with complex numbers,

$$(-1)^{\frac{1}{2}}(-1)^{\frac{1}{2}} = i^2 = -1,$$

which is a contradiction. See Definition 9.1 for the definition of this mysterious letter i. There is also the fallacy that

$$\sqrt{(-1)^2} = (-1)^{\frac{2}{2}} = (-1)^1 = -1,$$

whereas the fact is that $\sqrt{(-1)^2} = \sqrt{1} = 1$. The issue is that $\frac{2}{2}$ initially has an even denominator. Overall, exponent rules should be not be blindly applied when a base is negative.

Problem 3.57. By **rationalizing the denominator** of a fraction, we refer to a process of finding the same rational number in fraction form, but with a rational denominator. For example,

$$\frac{a}{\sqrt{b}} = \frac{a\sqrt{b}}{\sqrt{b} \cdot \sqrt{b}} = \frac{a\sqrt{b}}{b}.$$

Completing such a process can allow for more convenient arithmetic manipulations, such as finding common denominators. See if you can rationalize the denominators of

$$\frac{1}{\sqrt{2}+\sqrt{3}} \text{ and } \frac{1}{\sqrt{2}-\sqrt{3}+\sqrt{5}}$$

using the observation that

$$(\sqrt{a}-\sqrt{b})(\sqrt{a}+\sqrt{b}) = a-b.$$

Definition 3.58. Since an exponential function $f(x) = b^x$ is bijective onto $\mathbb{R}_+$ for $0 < b \neq 1$, we can define its inverse $g : \mathbb{R}_+ \to \mathbb{R}$, which is written as $g(y) = \log_b y$. This is called the **logarithm** with base b.

Similar to exponents, and often proven using exponent laws, there exist logarithmic laws.

Theorem 3.59. By the definition of inverses, we get our first two logarithmic identities for $0 < b \neq 1$:

1. For all $x \in \mathbb{R}$, it holds that $\log_b(b^x) = g(f(x)) = x$.
2. For all $y \in \mathbb{R}_+$, it holds that $b^{\log_b y} = f(g(y)) = y$.

These two rules combined tell us that $b^x = y$ if and only if $\log_b y = x$, with one direction of this biconditional statement being proven by each composition. Try showing it. Switching between exponential notation and logarithmic notation is an important technique.

Theorem 3.60. Let $b > 0$ and $c > 0$ be bases not equal to 1, $x > 0$ and $y > 0$ be arguments, and a be any real number. Then the following rules hold:

1. $\log_b 1 = 0$
2. $\log_b b = 1$
3. Product rule: $\log_b(xy) = \log_b x + \log_b y$
4. Quotient rule: $\log_b \left(\frac{x}{y}\right) = \log_b x - \log_b y$
5. Power Rule: $\log_b(x^a) = a \log_b x$
6. Change-of-base formula: $\log_b x = \frac{\log_c x}{\log_c b}$
7. Reciprocal rule: $\log_b c = \frac{1}{\log_c b}$
8. $\log_{(b^a)}(x^a) = \log_b x$
9. $\log_{\frac{1}{b}} x = -\log_b x$

Proof. The proofs largely involve converting the rules to exponential form, and using exponent rules. We will work backwards from the desired logarithmic identities while making sure that our steps are reversible, utilize the fact that exponential functions are bijective, and make repeated use of the identity $b^{\log_b y} = y$ that we stated in Theorem 3.59. We encourage the reader to try proving these identities independently before reading the brief proofs provided.

1. Converting to exponential form, $\log_b 1 = 0$ if and only if $b^0 = 1$, which is true.
2. Converting to exponential form, $\log_b b = 1$ if and only if $b^1 = b$, which is true.
3. Converting to exponential form, $\log_b(xy) = \log_b x + \log_b y$ if and only if

$$xy = b^{\log_b x + \log_b y} = b^{\log_b x} \cdot b^{\log_b y},$$

which is true.

4. Converting to exponential form, $\log_b\left(\frac{x}{y}\right) = \log_b x - \log_b y$ if and only if

$$\frac{x}{y} = b^{\log_b x - \log_b y} = \frac{b^{\log_b x}}{b^{\log_b y}},$$

which is true.

5. Converting to exponential form, $\log_b(x^a) = a\log_b x$ if and only if

$$x^a = b^{a\log_b x} = \left(b^{\log_b x}\right)^a,$$

which is true.

6. Clearing the denominator turns it into

$$\log_c b \log_b x = \log_c x.$$

Converting to exponential form yields

$$c^{\log_c b \log_b x} = x.$$

This is true because the left side reduces as

$$c^{\log_c b \log_b x} = \left(c^{\log_c b}\right)^{\log_b x} = b^{\log_b x} = x.$$

7. By the change-of-base formula,

$$\frac{1}{\log_c b} = \frac{\log_c c}{\log_c b} = \log_b c.$$

8. By the change-of-base formula and power rule,

$$\log_{b^a}(x^a) = \frac{\log_b(x^a)}{\log_b(b^a)} = \frac{a\log_b x}{a\log_b b} = \frac{a\log_b x}{a} = \log_b x.$$

9. By the change-of-base-formula,

$$\log_{\frac{1}{b}} x = \frac{\log_b x}{\log_b\left(\frac{1}{b}\right)} = \frac{\log_b x}{-1} = -\log_b x.$$

□

Corollary 3.61. The chain rule

$$\log_b x = \frac{\log_c x}{\log_c b}$$

can equivalently be written as

$$(\log_c b)(\log_b x) = \log_c x,$$

which makes it seem like the two b's cancel out. The generalization to a product

$$(\log_a b_1)(\log_{b_1} b_2)\cdots(\log_{b_n} x) = \log_a x$$

is called the **chain rule for logarithms**. As always, all of the bases must be positive (and not 1), and all of the inputs, otherwise known as arguments, must be positive.

Chapter 4

Indexed Objects

"Fundamental progress has to do with the reinterpretation of basic ideas."

– *Alfred North Whitehead*

"If somebody tells you a rule, break it. That's the only way to move things forward."

– *Hans Zimmer, MasterClass*

"It is in this gesture of "going beyond," to be something in oneself rather than the pawn of a consensus, the refusal to stay within a rigid circle that others have drawn around one - it is in this solitary act that one finds true creativity. All others things follow as a matter of course."

– *Alexander Grothendieck*

The intuition behind indexing is that if we have some distinguishable index cards, then we can write a (not necessarily unique) symbol on each card. In particular, if the index cards are stacked in an order, then we can speak of the symbol at a particular point or "index" in the order. The indexed objects that we will study are lists or tuples, sequences, series and products, some of which have variants with infinite entries or terms. We will end by studying a powerful method of reinterpreting nested sums and products, which we will call the "discrete Fubini's principle" [3]. The discrete Fubini's principle is commonly applied in combinatorics, especially for proving more complicated combinatorial identities.

4.1 Sequences

Definition 4.1. If I and S are non-empty sets and $f : I \to S$ is a function, then we define the following terms:

- f is an **indexing function** and may be called an **I-indexed family**
- I is the **indexing set** and its elements are called **indices**
- $\mathrm{Rng}(f)$ is the **indexed set** and its elements are called **entries**

In a context where the function f is treated as a tool for indexing, f_i denotes $f(i)$ and is called the **entry at index** i, and f is denoted by $(f_i)_{i\in I}$. Two indexing functions are said to be **equal** if they are equal as functions, meaning the indexing sets are the same and corresponding indices lead to the same output (Definition 2.24). There are two common specific types of indexing functions:

- If $I = [n]$ for some positive integer n, then f is called a **list** or **tuple** and may be denoted as $(f_i)_{i=1}^{n}$ or $(f_1, f_2, \ldots, f_n)$. More generally, the indexing set in a list may be taken to be any non-empty finite set, which is any set that is equipotent (meaning, in bijection) with $[n]$ for some positive integer n.
- If $I = \mathbb{Z}_+$, then f is called a **sequence** and may be denoted as $(f_i)_{i=1}^{\infty}$ or $(f_1, f_2, f_3, \ldots)$. More generally, the indexing set in a sequence may be taken to be any countably infinite set, which is any set that is equipotent with $\mathbb{Z}_+$.

The more general definitions of lists and sequences can be useful, such as with Fibonacci and Catalan numbers because the first index of each is 0 instead of 1; these two sequences will be explored in Volume 2.

We have defined lists and sequences rather formally. The reader would be well-advised to not worry about the formal definition in most cases, and instead simply think of them as finite or countably infinite sets written down in a particular order from left to right.

Definition 4.2. Though it is not a requirement for an indexing function, the indexing set might have a total order (and so the associated strict total order) on it (see Definition 5.39). For example, there is the standard order on $\mathbb{R}$ that is inherited by $[n]$ and $\mathbb{Z}_+$. This can be beneficial because it places the outputs in that order. Suppose $f : I \to S$ is an indexing function. If the I and S have total orders on them (and so the associated strict total orders), then f is said to be **strictly increasing**, **strictly decreasing**, **strictly monotone**, **non-decreasing**, **non-increasing**, or **monotone** according to whether f satisfies that property as a function (see Definition 5.24).

Strictly monotone lists will appear in the proof of the Erdős-Szekeres theorem in Volume 2, and non-decreasing and strictly increasing tuples will be a part of the rearrangement inequality (Theorem 12.18) and Chebyshev's inequality (Corollary 12.19) when we study multivariable inequalities.

Definition 4.3. If $A : I \to S$ is an indexing function, where S is a set of sets, then we can denote the union of all sets that are elements of the indexed set by

$$\bigcup_{i\in I} A_i = \{a : \exists i \in I, a \in A_i\}.$$

Here, we used the convention $A_i = A(i)$.

We defined the Cartesian product of an ordered pair of sets in Definition 2.15. Equipped with the definition of a list of sets, we can generalize this concept.

Definition 4.4. Given a list of sets $(A_1, A_2, \ldots, A_n)$ for a positive integer n, we denote and define its **Cartesian product** to be the set

$$A_1 \times A_2 \times \cdots \times A_n = \{(a_1, a_2, \ldots, a_n) : a_1 \in A_1, a_2 \in A_2, \ldots, a_n \in A_n\}.$$

In words, it is the set of all n-tuples such that, for each index $i \in [n]$, the ith entry is from A_i. If all of the A_i are the same set A, then we can use the notation

$$A^n = \underbrace{A \times A \times \cdots \times A}_{n \text{ copies of } A}.$$

Example. If one or more of the component sets of a Cartesian product are $\varnothing$, then the whole Cartesian product is $\varnothing$ because no tuples can be created. This also makes sense by contrapositive because, if a Cartesian product is non-empty, then each of the component sets must contain at least one element.

An idea that we will now explore is that of counting the number of entries in a list without writing out every entry. That is, we will calculate the length of a list, given some description of the list, which is an important skill in practice. Inevitably, there will need to be some pattern in the list in order for this task to be feasible, and we will usually need to know the first and last entries.

Example 4.5. The pages (not individual sheets) of a book are numbered from 1 to 100. How many pages are there in between 10 and 40 inclusive? "Inclusive" means we include the pages 10 and 40 at the ends.

Solution. It is tempting to say that the answer is $40 - 10 = 30$ but this is incorrect. The list of pages is

$$(10, 11, 12, \ldots, 40).$$

We subtract 9 from each entry to turn it into the list

$$(1, 2, 3, \ldots, 31).$$

This means there are 31 pages from page 10 to page 40 inclusive. □

Problem 4.6. The pages of a book are numbered from 1 to 100. How many pages are there in between 10 and 40 exclusive? Exclusive means we do not include the pages 10 and 40 at the ends.

In Example 4.5 and Problem 4.6, we saw that it is easy to make an "off-by-one error." More importantly, we found a method for determining the length of a list. The method generalizes nicely: assuming the given list has distinct entries at each index, we can apply injections to the list (so that the entries at all indices remain distinct at each step) until the list is of the form $(1, 2, 3, \ldots, n)$. This will mean the number of entries is n. Examples of useful injections to keep in our repertoire are division or multiplication by a non-zero constant, and addition and subtraction by a constant (we will call these "list transformations"). Addition or subtraction can be thought of as translating the list, and multiplication or division by a non-zero constant can be interpreted as scaling the list.

Problem 4.7. Find the number of multiples of 3, which are strictly greater than 110 and strictly less than 1000.

Problem 4.8. Though we will define arithmetic and geometric sequences later in Section 7.1, now is an appropriate time for the reader to prove the following statements.

1. Let a and $d \neq 0$ be real numbers. Then there are $m - n + 1$ entries in the list
$$(a + nd, a + (n+1)d, \ldots, a + md).$$
2. Let $b \neq 0$ and $r \neq 0, \pm 1$ be real numbers. Then there are $m - n + 1$ entries in the list
$$(br^n, br^{n+1}, \ldots, br^m).$$

The reason for the restrictions on the constants is to ensure that the terms of the sequences are all distinct.

Problem 4.9. Find the number of perfect cubes (as in, cubes of integers), which are strictly greater than -1001 and strictly less than 1001.

4.2 Sums and Products

Definition 4.10. Finite and infinite series are defined as follows.

1. If n is a positive integer and $a : [n] \to \mathbb{R}$ is a list, then the corresponding **finite series** is denoted by and defined as
$$\sum_{i=1}^{n} a_i = a_1 + a_2 + \cdots + a_n.$$
If the indexing set is $\{p, p+1, p+2, \ldots, q\}$ for integers $p \leq q$, then the corresponding finite series can be denoted by
$$\sum_{i=p}^{q} a_i = a_p + a_{p+1} + \cdots + a_q.$$
If the upper bound on the index in a sum is less than the lower bound, then this "empty sum" is defined to be 0; this convention can help with writing general formulas where there are edge cases.

   More generally, for any finite indexing set I, if $a : I \to \mathbb{R}$ is a list then the corresponding finite series is denoted by $\sum_{i \in I} a_i$, which denotes the sum of the outputs $a(i)$. This notation is acceptable because, thanks to the commutativity of addition, the order in which the a_i are added does not matter.
2. If $a : \mathbb{Z}_+ \to \mathbb{R}$ is a sequence, then the corresponding **infinite series** is denoted by and defined as
$$\sum_{i=1}^{\infty} a_i = a_1 + a_2 + a_3 + \cdots = \lim_{n \to \infty} \sum_{i=1}^{n} a_i.$$

For each positive integer n, the term $\sum_{i=1}^{n} a_i$ is called the nth **partial sum** of the series. The value of the infinite series is the value that the partial sums approach as n gets arbitrarily large. This idea of **convergence** can be made formal using limits from calculus. If the indexing set is $\{p, p+1, p+2, \ldots\}$ for some integer p, then the corresponding infinite series is

$$\sum_{i=p}^{\infty} a_i = a_p + a_{p+1} + a_{p+2} + \cdots = \lim_{n\to\infty} \sum_{i=p}^{n} a_i,$$

and for each positive integer n, its nth partial sum is $\sum_{i=p}^{p+n-1} a_i$.

Unlike a finite series, if J is a countably infinite set and $a : J \to \mathbb{R}$ is a sequence, then the expression $\sum_{j\in J} a_j$ is not necessarily well-defined. This is because rearranging the terms of an infinite series does not necessarily lead to the same final value. Readers interested in analysis should look up the concepts of absolute convergence and conditional convergence, and the Riemann series theorem.

Both finite series and infinite series are called **series** or a **sum** for short. In a series, each **term** a_i is called a **summand**. The symbol $\sum$ in the context of a series is called **summation notation**.

Example. It is important to note that not all sequences sum to a convergent series. Here are some examples:

- Of course,
$$1 + 2 + 3 + 4 + \cdots$$
does not converge and instead goes to infinity because the partial sums $1 + 2 + \cdots + n$ get arbitrarily large. Such a series is called **divergent**.

- Even though the individual terms of the **harmonic series**
$$1 + \frac{1}{2} + \frac{1}{3} + \frac{1}{4} + \cdots$$
get arbitrarily close to 0, the partial sums (surprisingly!) get arbitrarily large. See Example 4.11.

- The partial sums of **Grandi's series**
$$1 + (-1) + 1 + (-1) + \cdots$$
oscillate between 1 ad 0, so it does not converge.

Example 4.11. Show that partial sums of the harmonic series $1 + \frac{1}{2} + \frac{1}{3} + \frac{1}{4} + \cdots$ approach infinity as $n \to \infty$.

Solution. We will aim to get a lower bound on the mth partial sum of the harmonic series. And that lower bound needs to go to infinity as n goes to infinity. The crucial observation is that we should look at the harmonic series in the chunks that exist in between terms whose indices are powers of 2. For example,

$$\begin{aligned} 1 + \frac{1}{2} + \left(\frac{1}{3} + \frac{1}{4}\right) + \left(\frac{1}{5} + \frac{1}{6} + \frac{1}{7} + \frac{1}{8}\right) &> \frac{1}{2} + \frac{1}{2} + \left(\frac{1}{4} + \frac{1}{4}\right) + \left(\frac{1}{8} + \frac{1}{8} + \frac{1}{8} + \frac{1}{8}\right) \\ &= \frac{1}{2} + \frac{1}{2} + 2 \cdot \frac{1}{4} + 4 \cdot \frac{1}{8} \\ &= \frac{1}{2} + \frac{1}{2} + \frac{1}{2} + \frac{1}{2}. \end{aligned}$$

In general, for each positive integer $n \geq 2$,

$$\sum_{k=2^{n-1}+1}^{2^n} \frac{1}{k} > \sum_{k=2^{n-1}+1}^{2^n} \frac{1}{2^n} = (2^n - 2^{n-1}) \cdot \frac{1}{2^n} = 2^{n-1} \cdot \frac{1}{2^n} = \frac{1}{2}.$$

Thus, for each positive integer n,

$$\sum_{k=1}^{2^n} \frac{1}{k} = 1 + \sum_{i=0}^{n-1} \sum_{k=2^i+1}^{2^{i+1}} \frac{1}{k} > \frac{1}{2} + n \cdot \frac{1}{2} = \frac{n+1}{2}.$$

Since $\frac{n+1}{2}$ grows unboundedly as $n \to \infty$ and since the partial sums of the harmonic series are bounded below by this, the partial sums go to infinity as $n \to \infty$. □

Problem 4.12. The sum $\sum_{i=s}^{t} a_i$ starts with $i = s$ and then i goes up by increments of 1 up to and including $i = t$. Find a way of expressing (in summation notation) the sum of the same a_i but in the other direction, from $i = t$ to $i = s$. Remember, the counter or indexing variable can only go up by 1's, not down and not by any other increment.

Definition 4.13. The multiplicative analogue of summation notation is **product notation**. The symbol $\sum$ for finite and infinite series is replaced by $\prod$ for finite and infinite **products**. For positive integers n, finite sets I, and integers $p \leq q$, the meanings of

$$\prod_{i=1}^{n} a_i, \ \prod_{i=p}^{q} a_i, \ \prod_{i \in I} a_i, \ \prod_{i=1}^{\infty} a_i, \ \prod_{i=p}^{\infty} a_i$$

are analogous to the definitions of $\sum$ notation, where addition is replaced by multiplication. The multiplicative analogue of a summand is a **multiplicand**. For each positive integer n, just like how the nth partial sum of an infinite series is the sum of the first n summands, the nth **partial product** of an infinite product is the product of the first n multiplicands.

Definition 4.14. The empty sum is defined as the additive identity $\sum_{i \in \varnothing} a_i = 0$ and the empty product is defined as the multiplicative identity $\prod_{i \in \varnothing} a_i = 1$. This convention is useful at times.

Definition 4.15. There are many scenarios in which there are **nested** sums or products (possibly both), which means the summands or multiplicands themselves contain sums or products; this nesting can have any finite depth. Sums and products in one nesting should never share symbols for indices. When writing or interpreting a nesting, one should go from left to right and, at each step into the nesting, consider each appearance of the symbols for indices from previous levels to be a constant. For example,

$$\prod_{i \in I} \sum_{j \in J} a_i b_j = \prod_{i \in I} \left(a_i \sum_{j \in J} b_j \right) = \left(\prod_{i \in I} a_i \right) \cdot \left(\sum_{j \in J} b_j \right)^{|I|},$$

where $|I|$ denotes the number of elements in I. We were able to factor a_i out of $\sum_{j \in J} a_i b_j$ in the first step using the distributive law because i is a constant at that depth. For those with experience with basic programming, this iterative mechanism of summation and product notation is akin to loops, with nested sums and products being like nested loops. To avoid ambiguity, it is possible to use parentheses like

$$\prod_{i \in I} \sum_{j \in J} a_i b_j = \prod_{i \in I} \left(\sum_{j \in J} (a_i b_j) \right)$$

to clarify what lies in the summands or multiplicands, but this is not always necessary.

Definition 4.16. If the indexing set is

$$I = \{(i, j) \in [n] \times [n] : i < j\}$$

for some positive integer n, then two other ways of writing the sum $\sum_{(i,j)\in I} a_{i,j}$ are

$$\sum_{1\leq i<j\leq n} a_{i,j} = \sum_{i=1}^{n-1} \sum_{j=i+1}^{n} a_{i,j}.$$

Both deserve some commentary:

- The left side is an example of summation notation where it is not clear which symbols represent the indices, and even if we knew them, it has not been made clear to what set they belong. In such cases, the convention is to determine which symbols have a fixed value (such as n), assume that the other symbols are all indices (such as i and j), and iterate over all integer values of these presumed indices that satisfy all other restrictions (such as the fact that i and j are integers that fall in $[1, n]$ and satisfy $i < j$). We will see much of this when studying Vieta's formulas (Theorem 11.43) and the principle of inclusion-exclusion in Volume 2.

- The right side is an example of writing the sum in a way that gives explicit instructions on how to perform the iteration, like a computer program. It is not always feasible, nor is it always necessary. Moreover, it is an example of a nested summation where the range of an inner index depends on an outer index, since j begins at $i + 1$.

When there are multiple indices like i and j in the subscript, $a_{i,j}$ may be written without the comma as a_{ij} if there is no chance of mistaking ij as a multiplication or concatenation.

Theorem 4.17. There are a few ways in which sums and products are commonly manipulated. The basic variants are stated below, and we expect that the reader can adapt them to other circumstances, such as the index starting at an integer other than 1, as needed. If I and J are finite indexing sets, then:

$$\sum_{i\in I} (a_i + b_i) = \sum_{i\in I} a_i + \sum_{i\in I} b_i,$$

$$\prod_{i\in I} a_i b_i = \left(\prod_{i\in I} a_i\right) \cdot \left(\prod_{i\in I} b_i\right)$$

$$\sum_{i\in I} c a_i = c \cdot \sum_{i\in I} a_i,$$

$$\prod_{i\in I} c a_i = c^{|I|} \cdot \prod_{i\in I} a_i$$

$$\sum_{i\in I} \sum_{j\in J} a_i b_j = \sum_{i\in I} \left(a_i \sum_{j\in J} b_j \right) = \left(\sum_{i\in I} a_i\right) \cdot \left(\sum_{j\in J} b_j\right)$$

We state these without proof as they are straightforward consequences of the rules of arithmetic. There are many other ways in which summations and products may be manipulated. We advise the reader to not memorize these rules, but rather to understand the meaning of the notation and apply the rules of arithmetic as seen fit.

Before reading the rest of this chapter, consider reviewing the definition of a matrix in Appendix A.

Theorem 4.18 (Discrete Fubini's principle)**.** The essence of this principle is that, if there are nested summations, then it may be possible to alter the order in which the nesting occurs. This cannot be done blindly because there are numerous ways in which the indexing variables might depend on each other. While we are not aware of one general result that captures all possibilities, the results below are some of the most common instances of this idea, and the method used to prove them can be modified to suit other scenarios. Let m and n be positive integers and let a_{ij} be a real number for every $(i,j) \in [m] \times [n]$. Then

$$\sum_{i=1}^{m}\sum_{j=1}^{n} a_{ij} = \sum_{j=1}^{n}\sum_{i=1}^{m} a_{ij}.$$

Proof. We are tempted to provide the following "proof without words."

$$\begin{bmatrix} a_{11} & a_{12} & a_{13} & \cdots & a_{1n} \\ a_{21} & a_{22} & a_{23} & \cdots & a_{2n} \\ a_{31} & a_{32} & a_{33} & \cdots & a_{3n} \\ \vdots & \vdots & \vdots & \ddots & \vdots \\ a_{m1} & a_{m2} & a_{m3} & \cdots & a_{mn} \end{bmatrix}$$

As an explanation, the idea is to place the a_{ij} from the sum in an $m \times n$ matrix that has m rows and n columns, where a_{ij} occupies the entry at the intersection of row i and column j. Then one way to find the sum of the a_{ij} is to add the sum of all rows, and another way to find the sum of the a_{ij} is to add the sum of the all columns. The two methods yield the same final sum, so the two expressions are equal. □

Theorem 4.19 (Discrete Fubini's principle)**.** The same method as in Theorem 4.18 works for triangular arrays in square matrices, as shown below.

$$\begin{bmatrix} a_{11} & a_{12} & a_{13} & \cdots & a_{1n} \\ & a_{22} & a_{23} & \cdots & a_{2n} \\ & & a_{33} & \cdots & a_{3n} \\ & & & \ddots & \vdots \\ & & & & a_{nn} \end{bmatrix} \longrightarrow \sum_{i=1}^{n}\sum_{j=i}^{n} a_{ij} = \sum_{j=1}^{n}\sum_{i=1}^{j} a_{ij},$$

$$\begin{bmatrix} a_{11} & \cdots & a_{1(n-2)} & a_{1(n-1)} & a_{1n} \\ a_{21} & \cdots & a_{2(n-2)} & a_{2(n-1)} & \\ a_{31} & \cdots & a_{3(n-2)} & & \\ \vdots & .\cdot^{\cdot} & & & \\ a_{n1} & & & & \end{bmatrix} \longrightarrow \sum_{i=1}^{n}\sum_{j=1}^{n+1-i} a_{ij} = \sum_{j=1}^{n}\sum_{i=1}^{n+1-j} a_{ij},$$

$$\begin{bmatrix} a_{11} & & & & \\ a_{21} & a_{22} & & & \\ a_{31} & a_{32} & a_{33} & & \\ \vdots & \vdots & \vdots & \ddots & \\ a_{n1} & a_{n2} & a_{n3} & \cdots & a_{nn} \end{bmatrix} \longrightarrow \sum_{i=1}^{n}\sum_{j=1}^{i} a_{ij} = \sum_{j=1}^{n}\sum_{i=j}^{n} a_{ij},$$

$$\begin{bmatrix} & & & & a_{1n} \\ & & & a_{2(n-1)} & a_{2n} \\ & & a_{3(n-2)} & a_{3(n-1)} & a_{3n} \\ & .\cdot^{\cdot} & \vdots & \vdots & \vdots \\ a_{n1} & \cdots & a_{n(n-2)} & a_{n(n-1)} & a_{nn} \end{bmatrix} \longrightarrow \sum_{i=1}^{n}\sum_{j=n+1-i}^{n} a_{ij} = \sum_{j=1}^{n}\sum_{i=n+1-j}^{n} a_{ij}.$$

The same results hold if we replace all of the summations $\sum$ with products $\prod$ in the variations of the discrete Fubini's principle. There is no need to memorize these identities because they follow immediately from the proof technique shown below.

Problem 4.20. Let n be a positive integer and let a_{ij} be a real number for every $(i,j) \in [n] \times [n]$ such that $i < j$. Prove that

$$\sum_{i=1}^{n-1}\sum_{j=i+1}^{n} a_{ij} = \sum_{j=2}^{n}\sum_{i=1}^{j-1} a_{ij}.$$

Chapter 5

Equality and Order

"An equation has no meaning for me unless it expresses a thought of God."

– *Srinivasa Ramanujan*

"For an analyst, an equation is just two inequalities."

– *Anonymous*

Equations and inequalities are the two most effective ways of relating real numbers to each other. These relations not only take on the form of static propositions, but also dynamic predicates where we want to find all the values of variables that make the equation or inequality true. This act, known as finding the "solutions" to the predicate, is a core aspect of mathematical activity. We will look at some common methods of solving equations and inequalities, while observing that the technique underlying all of them is expansion-and-contraction: find a possibly larger set of candidates and eliminate anything that does not work. Finally, we will take a look at abstract orders, including total orders, partial orders, and well-orders, with a nod towards the fact that antisymmetry in a partial order is one of the most powerful tools lurking in the background of mathematics.

5.1 Equations

Definition 5.1. We previously defined an equation as the assertion that two objects are the same (Definition 2.20). An **equation** can also refer to a predicate like

$$A(x_1, x_2, \ldots, x_n) = B(x_1, x_2, \ldots, x_n),$$

where A and B are expressions of some sort, usually an amalgam of functions and operations acting on the variables x_i. Not all assignments of the x_i necessarily result in a true equation. If a tuple $(x_1, x_2, \ldots, x_n)$ of fixed values causes this equation to be true, then we say that the tuple **satisfies** the equation. Finding all such tuples is called **solving** the equation. If all possible tuples within a certain domain satisfy the equation, then the equation is said to be an **identity** on that domain.

Example. An example of an ordinary equation with specified objects is $1 = 1$. An example of a predicate equation is $x + 1 = 2x$. Its only solution is the number 1, so we say that $x = 1$ is the only solution. An example of an identity that holds for all $(x, y) \in \mathbb{R} \times \mathbb{R}$ (and, in fact, for complex numbers too) is the **difference of squares** factorization

$$x^2 - y^2 = (x - y)(x + y).$$

Definition 5.2. The simplest equations are **linear equations** which have sums of variables times constants, called **coefficients**, equal to a constant. These equations look like

$$a_1x_1 + a_2x_2 + \cdots + a_nx_n = b,$$

where the x_i are the variables, the a_i are the coefficients, and b is the constant.

Example 5.3 (One-variable linear equation)**.** As a simple example, find all real numbers x such that $ax + b = 0$, where b is a real constant and a is a non-zero real constant.

Solution. Suppose x is a real number that is a solution to the equation. Then $ax + b = 0$. First we subtract b from both sides (this means applying the function $f(t) = +(t, -b)$ to both sides) to get

$$ax = -b.$$

Then we divide both sides by the non-zero real a (this means applying the function $g(t) = \times(t, a^{-1})$ to both sides) to get

$$x = -\frac{b}{a}.$$

So, the only possible solution is this number. However, we need to verify that it actually works. If $x = -\frac{b}{a}$, then

$$ax + b = a \cdot \left(-\frac{b}{a}\right) + b = -b + b = 0,$$

so it does work. We can either substitute in all solutions found like this to check which ones work, or we can check that all steps were reversible in the original "only if" direction. This check is needed because the "only if" direction merely finds a *superset* of the set of solutions, but not all of those values are necessarily solutions. Technically, what we have done here is use the fact that, if $P(x)$ and $Q(x)$ are predicates such that

$$\forall x \in \mathbb{R}, (P(x) \implies Q(x))$$

and $P(c)$ is true for some real constant c, then $Q(c)$ is true. This produces the subset relation

$$S = \{x \in \mathbb{R} : P(x)\} \subseteq \{x \in \mathbb{R} : Q(x)\} = T,$$

since

$$c \in S \iff P(c) \implies Q(c) \iff c \in T.$$

This can be conceptualized as expanding to a larger set of potential solutions and then contracting back to the smaller set of actual solutions. Ironically, the larger set might be easier to express. In physical situations and mathematical scenarios, there might be elements of the larger set that are not elements of the smaller set, say due to an extra restriction. In such cases, these elements are called **extraneous solutions**. □

Problem 5.4. As we saw in the solution to Example 5.3, a standard technique used to solve an equation is to apply the same function to both sides of the equation. Often, we apply a binary operation with the first entry as a side of the equation and the second entry as some specific number, which makes it equivalent to apply a univariate function to both sides. Although equalities are preserved under such an application of a function, show that an inequation like $a \neq b$ is not necessarily preserved. A specific example will suffice.

Definition 5.5. A **system of equations** is a finite collection of equations. A **solution** to the system is a tuple of values for the variables such that the tuple of values satisfies every equation; we typically wish to find *all* solutions, though this is not always feasible and we might count ourselves lucky to find any of the solutions at all.

Barring the theory of linear algebra, there are three main elementary techniques for finding all solutions to a system of linear equations: substitution, elimination, and what we call "smashing." All of the techniques rely on the expansion-contraction method: find a superset of the solutions in the "only if" direction and then remove any that do not actually work in the "if" direction.

- **Substitution**: Isolate a variable in one equation by writing it as an expression in terms of the other variables in the equation. Then remove this variable from the other equations by substituting the expression into all of the other equations in which the variable appears.

- **Elimination**: Given two equations $a = b$ and $c = d$, a third equation that is true is $ta + sc = tb + sd$ for any real s, t. By adding a suitable multiple of one equation to a suitable multiple of another equation, we might be able to simplify the system, say by removing a variable. This is a modification of Theorem 3.16.

- **Smashing**: Add or multiply together all or most of the given equations. This works because if $a = b$ and $c = d$, then $a + c = b + d$ and $ac = bd$. Again, the engine driving this is Theorem 3.16.

We will see another technique for solving equations that utilizes symmetry to induce an order when we study inequalities (see Theorem 5.38).

Example 5.6 (Substitution)**.** Solve this system of equations:

$$x^2 + y^2 = 1,$$
$$x = y.$$

Solution. Suppose (x, y) is a solution to both equations. Thanks to the second equation, we can substitute x in for y in the first equation. This yields

$$2x^2 = 1 \iff x^2 = \frac{1}{2} \iff x = \pm\frac{1}{\sqrt{2}},$$

and y has the same value as x. Since the steps are reversible, the only two solutions are

$$x = y = \frac{1}{\sqrt{2}}, \text{ or } x = y = -\frac{1}{\sqrt{2}}.$$

By "the only," we mean that they are both solutions and there are no other solutions. □

Example 5.7 (Elimination)**.** Solve this system of equations:

$$4a + 3b = 90,$$
$$3a + 4b = 85.$$

Solution. We will use Theorem 3.16 several times in order to add and subtract equations together. Adding the two equations yields

$$7a + 7b = 175 \implies a + b = 25.$$

Subtracting the two equations yields

$$a - b = 5.$$

Adding the two new equations yields

$$2a = 30 \implies a = 15.$$

Subtracting the two new equations yields

$$2b = 20 \implies b = 10.$$

Substituting $(a, b) = (15, 10)$ into the original two equations shows that both equations are satisfied, so $(15, 10)$ is the only solution. □

Problem 5.8. Let p, q, r be real constants. Suppose x, y, z are non-zero real numbers that solve the following system of equations:

$$p = x + \frac{1}{y},$$
$$q = y + \frac{1}{z},$$
$$r = z + \frac{1}{x}.$$

Express $xyz + \frac{1}{xyz}$ in terms of p, q, r using arithmetic operations.

Problem 5.9. Find all real numbers a such that a is its own additive inverse. Similarly, find all real numbers b such that b is its own multiplicative inverse.

Problem 5.10. It is important to ensure that each step of manipulating an equation makes sense and does not accidentally make an illegal move. Observe the following "proof" that starts with any two equal numbers and ends with $2 = 1$:

$$\begin{aligned} a &= b \\ a^2 &= ab \\ a^2 - b^2 &= ab - b^2 \\ (a-b)(a+b) &= (a-b)b \\ a+b &= b \\ 2a &= a \\ 2 &= 1. \end{aligned}$$

Can you spot the error?

Example 5.11. Find all solutions to the following system of equations:

$$\begin{aligned} 2x + 3y &= 4, \\ 6x + 9y &= 13. \end{aligned}$$

Solution. Suppose there exists a solution (x, y). Then the second equation implies that

$$2x + 3y = \frac{13}{3}.$$

The first equation says that

$$2x + 3y = 4.$$

Then $4 = \frac{13}{3}$, which contradicts that these two numbers are unequal. Thus, there are no solutions. □

Example 5.12. Find all solutions of the following system of equations:

$$\begin{aligned} 2x + 3y &= 4, \\ 4x + 6y &= 8. \end{aligned}$$

Solution. Suppose (x, y) is a solution to the system. The two equations imply each other because the second is twice the first and the first is half the second. We need only consider the first equation because the second equation provides no new information. We can isolate y in $2x + 3y = 4$ to get

$$y = \frac{4 - 2x}{3}.$$

So every solution is of the form $\left(t, \frac{4-2t}{3}\right)$ for some real t. The question is whether all elements of the set

$$\left\{\left(t, \frac{4-2t}{3}\right) : t \in \mathbb{R}\right\}$$

are solutions. We can check that this is the case by plugging these coordinates into $2x + 3y$, which yields

$$2x + 3y = 2t + 3\left(\frac{4-2t}{3}\right) = 2t + 4 - 2t = 4.$$

So, this is indeed the complete set of solutions. This way of expressing the set of all solutions in terms of one or more variables, like t, that have a high degree of freedom is called a **parametrization**, with t being called a **parameter**. □

5.2 Inequalities

The reader should be familiar with the notion of order on the real numbers. This allows us to produce inequalities like $2 \geq 1$ and $0 < 1$ that compare where the numbers lie in relation to each other on the number line. Let us review the properties of real number inequalities. Firstly, we will decide how to compare two real numbers in decimal representation.

Theorem 5.13. Two non-negative integers may be compared according to how many 1's are required to be added up to be equal to each. We will need this for comparing digits, as in $0, 1, 2, \ldots, 9$. For example,

$$3 = 1 + 1 + 1 > 1 + 1 = 2.$$

Now, let a and b be positive real numbers whose decimal representations are

$$a = a_n a_{n-1} \ldots a_1 a_0.a_{-1}a_{-2}a_{-3}\ldots,$$
$$b = b_m b_{m-1} \ldots b_1 b_0.b_{-1}b_{-2}b_{-3}\ldots,$$

where we choose decimal representations such that it is not true that all digits of sufficiently small indices are 9 (Volume 3 says that there always exists a unique representation that avoids this property of trailing 9's). If $n > m$ then $a > b$, and if $m > n$ then $b > a$. Now suppose $n = m$. Then we go through the digits from left to right and compare the digits of a, b that have corresponding indices. If there are no conflicting digits, then $a = b$. Otherwise, suppose $a_k \neq b_k$ are the first, meaning leftmost, digits at which a conflict occurs. If $a_k > b_k$ then $a > b$, and if $a_k < b_k$ then $a < b$, regardless of what happens at digits further to the right. This property of base representations makes comparing decimal representations easy. As for what happens if one or both of a, b are non-positive:

1. All negatives are less than all positives.

2. Zero is less than all positives and greater than all negatives.

3. If both a, b are negative then we can use the fact that $a > b$ if and only if $-b > -a$, which reduces the problem to comparing two positive reals.

We will skip providing a proof because it would require manipulating infinite series.

Definition 5.14. There is a difference between **strict** inequalities $<$ in which equality is not an option, and **non-strict** inequalities $\leq$ in which equality is possible. This is because the non-strict version $a \leq b$ is equivalent to "$a < b$ or $a = b$," and the strict version is equivalent to "$a \leq b$ and $a \neq b$." This is due to the trichotomy law (Theorem 5.16).

Example. The difference between strict and non-strict can have a giant effect effect in an argument, in terms of strength or validity, so one should pay attention to which one applies. A special case is $x \geq 0$ versus $x > 0$, where non-negative versus positive can make a world of difference.

Definition 5.15. An inequality I is **stronger** or **sharper** than an inequality J (and J is called **weaker**) if I is stronger as a conclusion, meaning I implies J.

Example. The strict inequality $a > b$ implies its non-strict counterpart $a \geq b$, so $a > b$ is stronger than $a \geq b$ (though we would still have to check that $a > b$ is, in fact, true). Note that if $a \geq b$ holds, it is not necessarily true that $a > b$ because $a = b$ might hold instead. On that note, $a = b$ implies $a \geq b$, but the converse does not necessarily hold because $a > b$ might be true instead.

Let us now observe some of the abstract properties of inequalities to which we have become accustomed through practice.

Theorem 5.16. Inequalities of real numbers satisfy many properties, the most important of which are as follows for all $a, b, c \in \mathbb{R}$:

1. Reflexivity and anti-reflexivity: $a \leq a$ is true but $a < a$ is false

2. Trichotomy law: exactly one of $a < b, a = b, b < a$ is true. As a result, $a \leq b$ and $a > b$ are negations, and $a \geq b$ and $a < b$ are negations.

3. Connexity: at least one of $a \leq b$ and $b \leq a$ is true

4. Antisymmetry: if both $a \leq b$ and $b \leq a$ are true, then $a = b$

5. Transitivity: If $a \leq b$ and $b \leq c$, then $a \leq c$; if one or both of $a \leq b$ or $b \leq c$ is replaced by its strict counterpart in the hypothesis, then $a \leq c$ can be replaced by its strict counterpart in the conclusion.

6. The compatibility rules:

   - Additive compatibility: If $a > b$, then $a + c > b + c$. It follows that if $a \geq b$, then $a + c \geq b + c$.
   - Multiplicative compatibility: If $a > b$ and $c > 0$, then $ac > bc$. It follows that if $a \geq b$ and c is non-negative, then $ac \geq bc$. If $c < 0$, then applying these results leads to flipping directions:

$$a > b \implies a(-c) > b(-c) \implies bc > ac,$$
$$a \geq b \implies a(-c) \geq b(-c) \implies bc \geq ac.$$

   Since subtraction is addition in disguise and division by a non-zero real is multiplication in disguise, they have analogues of the compatibility rules. We leave it to the reader to write them out.

Problem 5.17. Let a, b, c, d be real. As an analogue of Theorem 3.16 for inequalities, prove that we can "add" and "multiply" inequalities as follows:

- If $a > b$ and $c > d$, then $a + c > b + d$.
- If $a \geq b$ and $c > d$, then $a + c > b + d$.
- If $a \geq b$ and $c \geq d$, then $a + c \geq b + d$.
- If $a > b > 0$ and $c > d > 0$, then $ac > bd$.

Multiplication results that involve non-strict inequalities or non-positive numbers are left to the reader to discover in the same manner.

Theorem 5.18. If a and b are integers, then $a > b$ if and only if $a - 1 \geq b$. An equivalent form of the second proposition is $a \geq b + 1$. This result allows us to switch between a strict inequality and a non-strict inequality when we are dealing with integers.

Proof. Let a and b be integers such that $a > b$. Then $a - b > 0$. This means that $a - b$ is a positive integer, so $a - b \geq 1$, which is equivalent to $a - 1 \geq b$ or $a \geq b + 1$. Conversely, suppose $a - 1 \geq b$. Then $a > a - 1 \geq b$, so $a > b$ by transitivity. □

There are several concepts related to bounds listed below of which one should be aware.

Definition 5.19. Let S be a non-empty set of real numbers.

1. An **upper bound** on S is a real number a such that for all $x \in S$, it holds that $x \leq a$. A **lower bound** on S is a real number b such that for all $x \in S$, it holds that $x \geq b$. Upper or lower bounds do not depend on each other and it is possible that one or both do not exist for a given set.

2. A **maximum** of S is an upper bound on S that is also an element of S. A **minimum** is S is a lower bound on S that is also an element of S. Maxima and minima do not always exist, even if upper and lower bounds exist, but they always exist for finite sets. However, when a maximum or minimum does exist, it is unique (try proving this, or see the argument for uniqueness of minima in Theorem 2.58). The maximum is denoted by $\max(S)$ and the minimum is denoted by $\min(S)$. These functions are easily extended to non-set collections such as lists and multisets (see Volume 2 for their definitions).

3. A **supremum** of S is a minimum element of the set of upper bounds on S. An **infimum** of S is a maximum element of the set of lower bounds on S. There is a semi-secret, but extremely important property, of the real numbers called **Dedekind completeness**, which says that if S has an upper bound then S has a supremum, and if S has a lower bound then S has an infimum. Note that $\sup(S)$ and $\inf(S)$, as they are denoted, are not necessarily elements of S.

The properties of the max, min, sup, inf functions are too numerous to list here. For example, for any positive α,

$$\max\{\alpha x : x \in S\} = \alpha \cdot \max(S)$$

and for any negative β,

$$\max\{\beta x : x \in S\} = \beta \cdot \min(S).$$

Another example is that, if we define max and min as binary operations on ordered pairs, then they are commutative and associative; this will be helpful when analyzing the gcd and lcm functions in Volume 3. We will simply allow such properties to come up in practice.

Theorem 5.20 (Archimedean property)**.** For every real number r, there exists an integer n such that $n > r$.

Proof. Suppose, for the sake of contradiction, that the negation of the Archimedean property is true. This means that there exists a real number r such that, for all integers n, $n \leq r$. So, the integers are bounded above by r. By the Dedekind completeness of $\mathbb{R}$ there exists a least upper bound ℓ of $\mathbb{Z}$. The trick is to note that $\ell - 1$ is not an upper bound on $\mathbb{Z}$ because it is strictly less than the supremum ℓ. So, there exists an integer m such that $\ell - 1 < m$. But then $\ell < m + 1$. Since $m + 1$ is an integer, this contradicts the fact that ℓ is an upper bound on $\mathbb{Z}$. Thus, our initial assumption was wrong. Instead, the Archimedean property holds, making the integers "unbounded." □

Problem 5.21. Find an infinite sequence of positive rationals $q_1, q_2, q_3, \ldots$ that gets closer and closer to 0, but never reaches 0. Formally, the sequence should satisfy the following condition:

$$\forall \epsilon > 0, \exists N \in \mathbb{Z}_+, \forall n \in \mathbb{Z}_+ : n \geq N \implies 0 < q_n < \epsilon.$$

Definition 5.22. To effectively describe sets of real numbers, such as domains, ranges, and codomains, we introduce **interval notation**. Let a and b be real numbers. There are nine types of intervals:

- The **closed** interval $[a, b]$ is the set $\{x \in \mathbb{R} : a \leq x \leq b\}$.
- The **open** interval (a, b) is the set $\{x \in \mathbb{R} : a < x < b\}$.
- The interval $[a, b)$ is the set $\{x \in \mathbb{R} : a \leq x < b\}$.
- The interval $(a, b]$ is the set $\{x \in \mathbb{R} : a < x \leq b\}$.
- The interval $[a, \infty)$ is the set $\{x \in \mathbb{R} : a \leq x\}$.
- The interval (a, ∞) is the set $\{x \in \mathbb{R} : a < x\}$.
- The interval $(-\infty, a]$ is the set $\{x \in \mathbb{R} : x \leq a\}$.
- The interval $(-\infty, a)$ is the set $\{x \in \mathbb{R} : x < a\}$.
- The interval $(-\infty, \infty)$ is the set $\mathbb{R}$ of all real numbers.

Since intervals are sets, they are often combined in various ways using set operations, such as union, intersection, complement, and difference.

Problem 5.23. Find a set $S \subseteq \mathbb{R}$ such that S has an upper bound and a lower bound, but S does not have a maximum or minimum.

Definition 5.24. Let X be a non-empty subset of $\mathbb{R}$. Then the function $f : X \to \mathbb{R}$ is said to be:

- **Strictly increasing** if for all $x_1, x_2 \in X$, it holds that $x_1 < x_2 \implies f(x_1) < f(x_2)$.
- **Strictly decreasing** if for all $x_1, x_2 \in X$, it holds that $x_1 < x_2 \implies f(x_1) > f(x_2)$.
- **Non-decreasing** if for all $x_1, x_2 \in X$, it holds that $x_1 < x_2 \implies f(x_1) \leq f(x_2)$.
- **Non-increasing** if for all $x_1, x_2 \in X$, it holds that $x_1 < x_2 \implies f(x_1) \geq f(x_2)$.

Moreover, a function that is strictly increasing or strictly decreasing is called **strictly monotone**, and a function that is non-decreasing or non-increasing is called **monotone**.

Composing monotonic or strictly monotonic functions produces more such functions in ways that we leave it to the reader to discover. For example, composing two strictly decreasing functions produces a strictly increasing function, like a double negative turning into a positive.

Theorem 5.25. Let X be a subset of $\mathbb{R}$ and let $f : X \to \mathbb{R}$ be function. Let x_1 and x_2 be any elements of X. Then:

1. If f is strictly increasing or non-decreasing, then $f(x_1) < f(x_2) \implies x_1 < x_2$.
2. If f is strictly decreasing or non-increasing, then $f(x_1) < f(x_2) \implies x_1 > x_2$.

Proof. The two proofs are very similar.

1. Suppose f is strictly increasing or non-decreasing and that $f(x_1) < f(x_2)$. Suppose, for contradiction, that $x_1 \geq x_2$. Then $f(x_1) \geq f(x_2)$, which is a contradiction. So, $x_1 < x_2$.
2. Suppose f is strictly decreasing or non-increasing and that $f(x_1) < f(x_2)$. Suppose, for contradiction, that $x_1 \leq x_2$. Then $f(x_1) \geq f(x_2)$, which is a contradiction. So, $x_1 > x_2$.

□

Lemma 5.26. Strictly monotone functions, whether strictly increasing or strictly decreasing, are injective.

Proof. We will show the proof for strictly increasing functions. Suppose f is strictly increasing. To prove that f is injective, it suffices to show that

$$x_1 \neq x_2 \implies f(x_1) \neq f(x_2).$$

There are two cases, $x_1 > x_2$ or $x_1 < x_2$. These are handled as follows:

$$\begin{aligned} x_1 > x_2 &\implies f(x_1) > f(x_2) \implies f(x_1) \neq f(x_2), \\ x_1 < x_2 &\implies f(x_1) < f(x_2) \implies f(x_1) \neq f(x_2). \end{aligned}$$

We leave the analogous proof for strictly decreasing functions to the reader. □

Theorem 5.27. Let $f : X \to Y$ be a strictly monotone function where X is a subset of $\mathbb{R}$ and Y is the range of f. Then f is invertible and:

1. If f is strictly increasing, then f^{-1} inherits this property.
2. If f is strictly decreasing, then f^{-1} inherits this property.

Proof. By Lemma 5.26 and the fact that Y is the range of f, we get that f is bijective onto Y. This is equivalent to f having an inverse $f^{-1} : Y \to X$.
For the inheritance property, we will tackle each case separately. Suppose y_1 and y_2 are elements of Y such that $y_1 < y_2$ Let $f^{-1}(y_1) = x_1$ and $f^{-1}(y_2) = x_2$.

1. If f is strictly increasing, then

$$f(x_1) = y_1 < y_2 = f(x_2) \implies f^{-1}(y_1) = x_1 < x_2 = f^{-1}(y_2),$$

where we have used the fact that

$$f(x_1) < f(x_2) \implies x_1 < x_2$$

from Theorem 5.25. Thus, f^{-1} is also strictly increasing.

2. If f is strictly decreasing, then

$$f(x_1) = y_1 < y_2 = f(x_2) \implies f^{-1}(y_1) = x_1 > x_2 = f^{-1}(y_2),$$

where we have used the fact that

$$f(x_1) < f(x_2) \implies x_1 > x_2.$$

Thus, f^{-1} is also strictly decreasing.

This inheritance property does not extend to non-decreasing and non-increasing functions because they are not necessarily injective, and so such a function does not even necessarily have an inverse. □

Theorem 5.28. The following are examples of functions with properties of strict monotonicity:

1. The reciprocal function $f(x) = \dfrac{1}{x}$ is strictly decreasing on $(-\infty, 0)$ and it is also strictly decreasing on $(0, \infty)$, but the two are separate since the reciprocal of any positive real is greater than the reciprocal of any negative real.

2. For any odd positive integer n, $f(x) = x^n$ is strictly increasing on all of $\mathbb{R}$. As a consequence, by Theorem 5.27, its inverse function $g(x) = \sqrt[n]{x} = x^{\frac{1}{n}}$ is strictly increasing on all of $\mathbb{R}$.

3. For even positive integers n, $f(x) = x^n$ is strictly decreasing on $(-\infty, 0]$ and strictly increasing on $[0, \infty)$. As a consequence, the inverse function $g(x) = \sqrt[n]{x} = x^{\frac{1}{n}}$ is strictly increasing on $[0, \infty)$, which is the domain of g.

4. For real $b > 1$, $f(x) = b^x$ is strictly increasing on $\mathbb{R}$. As such, its inverse $g(x) = \log_b x$ is also strictly increasing on $\mathbb{R}_+$, which the domain of g.

5. For real $0 < b < 1$, $f(x) = b^x$ is strictly decreasing on $\mathbb{R}$. As such, its inverse $g(x) = \log_b x$ is also strictly decreasing on $\mathbb{R}_+$, which is the domain of g.

Note that, since the exponential function b^x (for positive $b \neq 1$) is strictly monotonic, it is injective. Thus, we can "equate exponents" in the equation $b^x = b^y$ to get $x = y$.

Proof. The first property is a result of multiplicative compatibility as we just have to manipulate the inequality $x_1 < x_2$ into $\dfrac{1}{x_2} < \dfrac{1}{x_1}$. Rigorous proofs of the remaining properties follow from calculus, which is beyond the scope of the text. □

As a side remark, the compatibility rules for inequalities (Theorem 5.16) follow from applying the monotone functions $f(t) = +(t, c)$ and $g(t) = \times(t, c)$ to inequalities.

Lemma 5.29 (Equality lemmas). In inequalities, we often want to know if and when equality holds. There is a pair of lemmas that come in handy for this purpose:

1. If

$$a_1 \geq a_2 \geq \cdots \geq a_n,$$

then $a_1 = a_n$ if and only if

$$a_1 = a_2 = \cdots = a_n.$$

2. If

$$a_1 \geq b_1, a_2 \geq b_2, \ldots, a_n \geq b_n,$$

then

$$a_1 + a_2 + \cdots + a_n \geq b_1 + b_2 + \cdots + b_n,$$

where equality holds if and only if

$$a_1 = b_1, a_2 = b_2, \ldots, a_n = b_n.$$

Proof. These two lemmas are useful for determining the equality cases of multivariable inequalities, such as those studied in Chapter 12.

1. If all of the a_k are equal, then clearly $a_1 = a_n$. Conversely, suppose $a_1 = a_n$. We know that, for each $k \in [n]$,

$$a_1 \geq a_k \geq a_n \implies a_1 = a_k = a_n,$$

where we have used antisymmetry. So all of the a_k are equal.

2. The inequality

$$a_1 + a_2 + \cdots + a_n \geq b_1 + b_2 + \cdots + b_n$$

holds by adding together all of the inequalities in the hypothesis. The tricky part is the equality criterion. If equality holds in all of the $a_k \geq b_k$, then adding the equations up yields

$$a_1 + a_2 + \cdots + a_n = b_1 + b_2 + \cdots + b_n.$$

Conversely, suppose this last equation holds. It can be rewritten as

$$(a_1 - b_1) + (a_2 - b_2) + \cdots + (a_n - b_n) = 0.$$

Each summand $a_k - b_k$ is non-negative by assumption. If any of the $a_k \geq b_k$ is strict then $a_k - b_k$ is positive, leading to the contradiction

$$(a_1 - b_1) + (a_2 - b_2) + \cdots + (a_n - b_n) > 0.$$

Thus, $a_k = b_k$ for all $k \in [n]$.

□

Problem 5.30. Let $(a_k)_{k=1}^n$ and $(b_k)_{k=1}^n$ be n-tuples of positive reals such that

$$a_1 \geq b_1, a_2 \geq b_2, \ldots, a_n \geq b_n.$$

Prove that

$$a_1 a_2 \cdots a_n \geq b_1 b_2 \cdots b_n,$$

where equality holds if and only if

$$a_1 = b_1, a_2 = b_2, \ldots, a_n = b_n.$$

Definition 5.31. Just like equations, **inequalities** can take the form of predicates, where the **solution** is the set of values of the variables that satisfy the inequality. When we study multivariable inequalities momentarily and again in Chapter 12, we will find inequalities that are **identities** in the sense that they hold for all real values of the variables (or some other large domain, like the positive reals).

Example 5.32. Solve $-2x + 3 > 0$ for all real x that satisfy it.

Solution. Suppose x is a real number that satisfies the inequality. Just like with equations, we apply the same operations to both sides until we isolate x:

$$-2x + 3 > 0 \iff -2x > -3 \iff x < \frac{3}{2}.$$

The steps are reversible, so the set of solutions is $\left(-\infty, \frac{3}{2}\right)$. Note that dividing both sides by -2 flipped the inequality sign because $f(t) = \times(t, (-2)^{-1})$ is a decreasing function due to $(-2)^{-1}$ being negative. In general, one has to be careful when multiplying or dividing by a negative real number, or a variable that might store a negative value. □

Example 5.33 (Sign analysis)**.** Find all real x such that

$$\frac{x+2}{x-3} < 4.$$

Solution. It would be ideal if we could clear the denominator, but we do not know its sign and so we do not know whether the inequality symbol would be flipped. Instead, we take the constant over to the other side and combine everything:

$$0 > \frac{x+2}{x-3} - 4 = \frac{x+2-4(x-3)}{x-3} = \frac{-3x+14}{x-3}.$$

Thus, we want to know the real numbers x for which $\frac{-3x+14}{x-3}$ is negative. We can figure this out using a technique called "sign analysis" where we look at the signs of the factors $-3x + 14$ and $x - 3$, since the sign of $\frac{-3x+14}{x-3}$ is the sign of $-3x + 14$ times the sign of $x - 3$. Neither factor can have a 0 sign because $-3x + 14 = 0$ would make the whole quotient 0 and $x - 3 = 0$ would cause division by 0.
Note that, if $(+)$ represents a positive sign and $(-)$ represents a negative sign, then

$$(+) \cdot (+) = (-) \cdot (-) = (+),$$
$$(+) \cdot (-) = (-) \cdot (+) = (-).$$

Since we want an overall negative sign, we only care about the cases where

$$-3x + 14 < 0 \text{ and } x - 3 > 0,$$
$$-3x + 14 > 0 \text{ and } x - 3 < 0.$$

This yields the two intervals

$$\{x \in \mathbb{R} : -3x + 14 < 0\} \cap \{x \in \mathbb{R} : x - 3 > 0\} = \left(\frac{14}{3}, \infty\right) \cap (3, \infty) = \left(\frac{14}{3}, \infty\right),$$
$$\{x \in \mathbb{R} : -3x + 14 > 0\} \cap \{x \in \mathbb{R} : x - 3 < 0\} = \left(-\infty, \frac{14}{3}\right) \cap (-\infty, 3) = (-\infty, 3).$$

This yields a final solution set of

$$(-\infty, 3) \cup \left(\frac{14}{3}, \infty\right)$$

Sign analysis becomes more difficult as the number of factors in the numerator and denominator increases, due to the extensive casework necessary. As such, we will develop a more efficient alternative called "interval analysis" in Example 11.40. □

The following is a result that is easy to overlook, especially because of its name, but it has powerful implications, some of which we will see when we study multivariable inequalities, such as the Cauchy-Schwarz inequality (Theorem 12.9).

Theorem 5.34 (Trivial inequality). The inequality $x^2 \geq 0$ is an identity on the real numbers, with equality holding if and only if $x = 0$.

Proof. This is easy to see by doing casework on the sign of x. If $x = 0$, then $x^2 = 0$. Regardless of whether x is positive or negative, $x^2 = x \cdot x$ is positive, and equality does not hold in these non-zero cases. □

Problem 5.35. Prove that

$$\frac{x^2 + y^2}{2} \geq xy$$

for all reals x, y, with equality holding if and only if $x = y$.

Theorem 5.36 (Bernoulli's inequality). If m is a non-negative integer and x is a real number such that $x \geq -1$, then

$$(1 + x)^m \geq 1 + mx.$$

If $x = -1$ and $m = 0$, then we use the convention that $0^0 = 1$. Moreover, equality holds if and only if $m = 0$ or $m = 1$ or $x = 0$.

Proof. First we will prove that the stated equality cases are indeed sufficient for equality

$$\begin{aligned} m = 0 &\implies (1+x)^m = (1+x)^0 = 1 = 1 + 0 \cdot x = 1 + mx, \\ m = 1 &\implies (1+x)^m = (1+x)^1 = 1 + x = 1 + 1 \cdot x = 1 + mx, \\ x = 0 &\implies (1+x)^m = (1+0)^m = 1 = 1 + m \cdot 0 = 1 + mx. \end{aligned}$$

Now suppose $x \neq 0$. We will prove, by induction on $m \geq 2$, that the strict inequality

$$(1 + x)^m > 1 + mx$$

holds. In the base case $m = 2$, we want to prove that

$$(1 + x)^2 > 1 + 2x.$$

Upon expanding the left side and cancelling terms common to both sides, this becomes equivalent to $x^2 > 0$, which is true by the trivial inequality because $x \neq 0$. Now suppose there exists an integer $m \geq 2$ such that

$$(1 + x)^m > 1 + mx.$$

Then multiplying both sides of the induction hypothesis by $1 + x$ (recall that this is positive) yields

$$\begin{aligned} (1+x)^{m+1} &\geq (1+mx)(1+x) \\ &= 1 + x + mx + mx^2 \\ &= 1 + (m+1)x + mx^2 \\ &> 1 + (m+1)x. \end{aligned}$$

Thus, we have proven Bernoulli's inequality by induction. □

Corollary 5.37. Bernoulli's inequality might seem like a passing curiosity, but it actually does not live in a vacuum. It implies that for any positive real number r, positive integer t, and integer $n \geq 2$,

$$\left(1 + \frac{r}{n}\right)^{nt} > 1 + nt \cdot \frac{r}{n} = 1 + rt.$$

For those familiar with the basics of investing, this shows that compound interest at a rate of r is a strictly better deal than simple interest at a rate of r, as long as the number of times n that the compound interest is applied in a single time period is at least 2.

Proof. The formula for compound interest is $p\left(1+\frac{r}{n}\right)^{nt}$ and the formula for simple interest is $p(1+rt)$. Here, p is the principal (your original amount of money), r is the interest rate as a decimal like $0.05 = 5\%$, t is the number of time periods elapsed (this is usually the number of years), and n is the number of times that the compound interest is applied per time period (if one time period is a year, then examples of n could be 4 for quarterly, 12 for monthly, and 365 for daily compounding). On a related note, see Example 12.5 for why compounding more times leads to a better deal for the receiver than compounding fewer times. □

Theorem 5.38. A technique involving orders that will occasionally be useful in solving inequalities or equations or proving identities is symmetry.

1. For solving inequalities or equations, if, for each pair of variables out of some set of variables, pairwise swapping them does not alter the truth of the statement, then we can assume that that set of variables fall in a certain order. After finding solutions under this order, all other solutions can be found by finding all permutations of these solutions.

2. For proving multivariable inequalities (over a large domain) that are symmetric in some variables, the story is simpler. It simply suffices to assume a certain order of the symmetric variables and prove that the inequality holds in that case, and this proof will translate over to all other orders. For example, we will see Schur's inequality (Theorem 12.14), which says that, for all non-negative real numbers x, y, z and all positive real numbers t, it holds that

$$x^t(x-y)(x-z) + y^t(y-z)(y-x) + z^t(z-x)(z-y) \geq 0.$$

Since x, y, z are symmetric here, we can and will assume without loss of generality that $x \geq y \geq z$ in the proof. The same idea is used in the proof of Popoviciu's inequality (Theorem 12.28).

If two variables with symmetry are known to represent unequal numbers, we can even assume a strict inequality between them.

5.3 Abstract Orders

The inequality relation on real numbers can be generalized. Given a set, we might want to place its elements in a particular order. This idea of ordering the elements of a set is formalized as follows. For the upcoming definitions, it will be helpful to recall from Definition 2.16 that a binary relation on S is a subset of $S \times S$.

Definition 5.39. A **total order** on a set is a binary relation $\leq$ on the set that behaves very similarly to non-strict inequalities on real numbers. To be precise, $\leq$ is a total order on the set S if for all $a, b, c \in S$ the following properties hold:

- Connexity: $a \leq b$ or $b \leq a$
- Antisymmetry: if $a \leq b$ and $b \leq a$, then $a = b$
- Transitivity: if $a \leq b$ and $b \leq c$, then $a \leq c$

The associated **strict total order** $<$ is a binary relation on S that is defined as: $a < b$ holds if it is not true that $b \leq a$. Equivalently, $a < b$ holds if $a \leq b$ and $a \neq b$. The strict total order satisfies the following properties for all $a, b, c \in S$:

- Irreflexivity: $a < a$ is false
- Asymmetry (different from antisymmetry): if $a < b$ is true, then $b < a$ is false
- Transitivity: if $a < b$ and $b < c$, then $a < c$
- Trichotomy law: exactly one of $a < b, a = b, b < a$ is true

Note that the trichotomy law implies irreflexivity and asymmetry because $a = a$ is always true, so we do not really need to point out the first two properties listed, other than for pedagogical purposes.

Example. The $\leq$ relation on the real numbers is a total order, and it induces the strict total order $<$ on the real numbers. Visually, it allows us to place the real numbers on a line in such a way that two real numbers a, b satisfy $a < b$ if and only if a is to the left of b.

Problem 5.40. Let us go in the reverse direction by using a strict total order to induce a total order. Prove that if $<$ is a transitive trichotomous binary relation, then the binary relation $\leq$ is a total order, where $\leq$ is defined as: $a \leq b$ if and only if $a < b$ or $a = b$.

In the context of manual combinatorics (i.e. counting elements of a finite set by listing them out), there is a particular total order that can be helpful, which we will define momentarily as the lexicographical order.

Definition 5.41. A collection of distinct symbols or letters is called an **alphabet**. A finite list of symbols of the alphabet is called a **word** or **string**; the symbols are ordered from right to left. The **length** of a string is the number of symbols in it.

Example. The **binary alphabet** is the set $\{0, 1\}$ and its strings are finite lists of 0's and 1's.

Definition 5.42. Suppose $\mathcal{A}$ is an alphabet with a total order $\geq$ on it with the corresponding strict total order $>$. We can define a total order, called **lexicographical order**, on the set of strings made of n symbols from $\mathcal{A}$, for a fixed positive integer n, as follows. Suppose there are strings

$$a = a_1 a_2 \cdots a_{n-1} a_n,$$
$$b = b_1 b_2 \cdots b_{n-1} b_n$$

such that $a \neq b$. Then there exists an index $1 \leq i \leq n$ where $a_i \neq b_i$. Let j be the smallest such index at which a and b differ. If $a_j > b_j$ then $a > b$, and if $a_j < b_j$ then $a < b$.

Example. Lexicographical order is used in dictionaries and encyclopedias so that there is a standard way to find entries according to their alphanumeric label. The way that we compared decimal representations of real numbers in Theorem 5.13 is also a variation of lexicographical order, albeit where the strings have countably infinite letters.

Definition 5.43. As we have defined it, lexicographical order tells us how to compare only strings that have the same number of symbols. Dictionaries extend this idea by defining a new "blank" letter that is smaller than every other letter, and adding exactly enough blanks to the right end of the smaller word so that the two words have the same number of letters. Then it is possible to compare them using lexicographical order. There is a different extension, called **shortlex order**, that is sometimes more convenient from an organizational perspective. It says that, given two strings of different length, the string with fewer letters is automatically smaller than the string with more letters. This is not so useful in a dictionary setting when we might not know the exact spelling of a word (and so the length of the word might be unknown, making it difficult to locate).

Example. If we follow the example of dictionaries, aaa comes before ab. In shortlex, ab comes before aaa.

Problem 5.44. The binary alphabet is $\{0, 1\}$ with the strict total order $0 < 1$. In shortlex order, list all binary alphabet strings that consist of one, two, or three symbols.

What if there are notions of ordering where antisymmetry and transitivity hold, but connexity does not? That is, maybe it is possible to preserve the essence of inequalities without it being possible to compare every pair of elements. This motivates the next definition.

Definition 5.45. A **partial order** on a set S is a binary relation $\leq$ on S such that for all $a, b, c \in S$ the following properties hold:

1. Reflexivity: $a \leq a$
2. Antisymmetry: if $a \leq b$ and $b \leq a$, then $a = b$
3. Transitivity: if $a \leq b$ and $b \leq c$, then $a \leq c$

Note that connexity implies reflexivity, so every total order is a partial order.

Example. Given a set S, the subset relation $\subseteq$ on its power set $\mathcal{P}(S)$ is a partial order. Below is what is known as *Hasse diagram* of the partial order structure of the subsets of $\{a, b, c\}$.

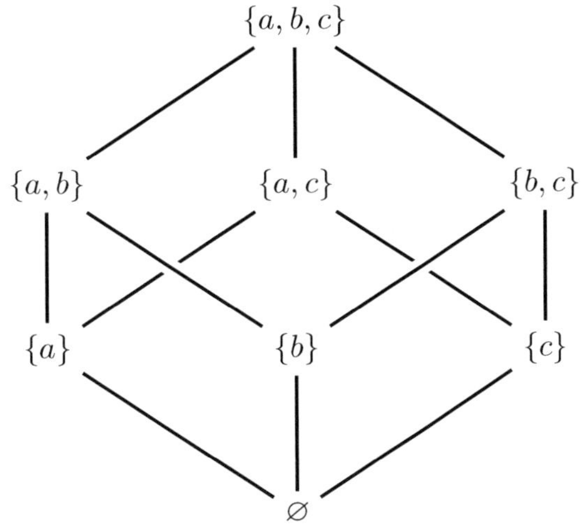

In Volume 3, we will define that a positive integer a divides a positive integer b if there exists a (necessarily positive) integer c such that $ac = b$. This relation is denoted by $a \mid b$, and it is an important partial order on the positive integers.

We cannot stress enough the incredible power of antisymmetry. It allows us to use two instances of the weaker notion of order to achieve an equality (equality is stronger than order because reflexivity says that if $a = b$ then $a \leq b$, but the other direction is not true in general). Antisymmetry will come up over and over again in our studies, especially with respect to numbers, subsets, and divisibility. For example, the "sandwiching"

$$a \leq b \leq a \implies a = b$$

will come up in probability, specifically with $b = 0$ and $b = 1$. In number theory, the antisymmetry of divisibility is an indispensable tool. In combinatorics and set theory, we have a variation of antisymmetry called the Schröder-Bernstein theorem. The list goes on.

Definition 5.46. One more kind of ordering, which we will not study abstractly, are well-orders. A **well-order** on a set S is a total order on S such that every non-empty subset of S has a least element under this ordering. More precisely, if the well-order is $\leq$ and T is a non-empty subset of S, then

$$\exists b \in T, \forall a \in T, b \leq a.$$

Example. The only example of a well-order that we will need is the ordinary order on $\mathbb{Z}_+$, or more generally, any subset of $\mathbb{Z}$ that has a lower bound. This is due to the well-ordering principle, which is described in Theorem 2.58 and used several times for studying number theory. Note that $\mathbb{Z}$ itself is not well-ordered because its elements descend downwards infinitely. Under the ordinary order, $\mathbb{R}$ is not well-ordered either, but it is one of the pathological consequences of the axiom of choice that a well-order can be defined on any set, including the reals.

We conclude by noting that partial orders include total orders, and total orders include well-orders, thus forming a nice hierarchy of these order types.

Chapter 6

Special Functions

"For fifteen days I struggled to prove that no functions analogous to those I have since called Fuchsian functions could exist; I was then very ignorant. Every day I sat down at my work table where I spent an hour or two; I tried a great number of combinations and arrived at no result. One evening, contrary to my custom, I took black coffee; I could not go to sleep; ideas swarmed up in clouds; I sensed them clashing until, to put it so, a pair would hook together to form a stable combination. By morning I had established the existence of a class of Fuchsian functions, those derived from the hypergeometric series. I had only to write up the results which took me a few hours."

– *Henri Poincaré, Science et Methode*

"A mathematician is a device for turning coffee into theorems."

– *Alfréd Rényi*

Having seen general properties of functions in Section 2.2, we will now see specific examples of functions that are common in mathematics. This includes absolute value, the signum function, floor function, and ceiling function. Absolute value naturally introduces us to the real triangle inequality, which has geometric and complex variants. The floor and ceiling functions are essential for rounding in a precise manner, and they have many properties, the most useful of which we will study.

6.1 Absolute Value

Definition 6.1. A particular definition of a function is said to be **piecewise** if the definition depends on different cases. To be clear, being piecewise is a property of the way a function is presented, not a property of a function itself. Nevertheless it is commonplace to speak of "piecewise functions" if there is a relevant definition of the function that is piecewise.

Example. A function might have different definitions on the intervals $(-\infty, 3]$, $(3, 17)$, and $[17, \infty)$. We denote such presentations by placing curly bracket on the left side of the cases as follows:

$$f(x) = \begin{cases} 2x^2 & \text{if } x \in (-\infty, 3] \\ 3x - 1 & \text{if } x \in (3, 17) \\ 5x^3 + 4x & \text{if } x \in [17, \infty) \end{cases}.$$

Some examples of piecewise functions that we will see are the absolute value and signum functions, and the floor and ceiling functions.

Recall, from Definition 3.29, that the absolute value of a real number is its distance from 0 on the real number line. So the absolute value function $|\cdot| : \mathbb{R} \to \mathbb{R}$ is defined by

$$|x| = \begin{cases} x & \text{if } x \geq 0 \\ -x & \text{if } x < 0 \end{cases}.$$

Theorem 6.2. There are several properties of the absolute value function that often come in handy. We state them here and leave the proofs to the reader because they are straightforward. Let x and y be real numbers and n be an integer. Then:

1. Instead of the given piecewise definition, an alternate unified definition of absolute value is $|x| = \sqrt{x^2}$, where we choose the non-negative square root as usual.

2. $|x| = 0$ if and only if $x = 0$, and otherwise $|x| > 0$. So, $|x| \geq 0$ always.

3. $|x| = \max(x, -x)$. Consequently, $|x| \geq x$, with equality holding if and only if $x \geq 0$.

4. $|xy| = |x| \cdot |y|$

5. For $x \neq 0$, $\left|\frac{1}{x}\right| = \frac{1}{|x|}$, and so $\left|\frac{y}{x}\right| = \frac{|y|}{|x|}$.

6. $|x^n| = |x|^n$, and, if n is an even integer, then $|x^n| = |x|^n = x^n$.

7. $|-x| = |x|$, and so $|x - y| = |y - x|$

8. Applying absolute value twice (or more times) is the same as applying it once, meaning $||x|| = |x|$, which is called its idempotent property.

Theorem 6.3. We often have to "open up" absolute values in equations or inequalities and do work on separate cases. There are three basic scenarios, for a real variable x and a non-negative constant c:

1. $|x| = c$ if and only if $x = \pm c$

2. $|x| > c$ if and only if $x > c$ or $x < -c$

3. $|x| < c$ if and only if $-c < x < -c$

We leave the proofs to the reader. One method is to square both sides (which is a reversible step here), take everything to one side, and use the difference of squares. The analogous statements for negative c are trivial or vacuous, and we encourage the reader to explore them.

Theorem 6.4 (Real triangle inequality)**.** For any real numbers x and y,

$$|x| + |y| \geq |x + y|.$$

Equality holds if and only if both of x and y are non-negative or both are non-positive. Another way of stating the equality condition that is sometimes useful is that $x = 0$ or $y = 0$ or $x = cy$ for some real $c > 0$.

Proof. We will work backwards from the given inequality using reversible steps. Since both sides of $|x|+|y| \geq |x + y|$ are non-negative, it is a reversible step to square both sides, which yields

$$\begin{aligned}
(|x| + |y|)^2 \geq |x + y|^2 &\iff |x|^2 + 2|x| \cdot |y| + |y|^2 \geq |x + y|^2 \\
&\iff x^2 + 2|xy| + y^2 \geq (x + y)^2 \\
&\iff x^2 + 2|xy| + y^2 \geq x^2 + 2xy + y^2 \\
&\iff |xy| \geq xy,
\end{aligned}$$

which is true. Equality holds if and only if $|xy| = xy$, which is true if and only if xy is non-negative. We know that xy is non-negative if and only if x and y are both non-negative or both non-positive. For the other way of stating the equality condition, if $x = 0$ or $y = 0$ then equality clearly holds. If neither x nor y is 0, then it must be hold that $\operatorname{sgn}(x) = \operatorname{sgn}(y)$, where the signum function is defined in Definition 6.7. Then we can simply define $c = \frac{x}{y}$, which must be positive due to the matching signs of x and y. □

Example 6.5 (Reverse triangle inequality)**.** Prove that, for all real x and y,

$$||x| - |y|| \leq |x - y|.$$

Solution. We could square the inequality as in the proof of the Triangle Inequality, and, in fact, this would make it easier to derive the equality criterion. However, since we are not asked to find the equality condition, let us practice using the triangle inequality instead. We get the two inequalities

$$|x| \leq |x-y| + |y|,$$
$$|y| \leq |y-x| + |x|.$$

These are equivalent to

$$|x| - |y| \leq |x-y|,$$
$$|y| - |x| \leq |x-y|.$$

Since $||x|-|y|| = \max(|x|-|y|, |y|-|x|)$, the desired inequality follows. □

Problem 6.6. Prove that, for any real numbers $x_1, x_2, \ldots, x_n$,

$$|x_1| + |x_2| + \cdots + |x_n| \geq |x_1 + x_2 + \cdots + x_n|.$$

Also prove that equality holds if and only if all the x_k are non-negative or all of them are non-positive.

Definition 6.7. The **signum function** gives the sign of a real number, meaning

$$\operatorname{sgn}(x) = \begin{cases} 1 & \text{if } x > 0 \\ 0 & \text{if } x = 0 \\ -1 & \text{if } x < 0 \end{cases}.$$

Theorem 6.8. The following properties hold for the signum function for real x and y:

1. $\operatorname{sgn}(-x) = -\operatorname{sgn}(x)$
2. $\operatorname{sgn}(x) = \dfrac{x}{|x|} = \dfrac{|x|}{x}$ for $x \neq 0$
3. $\operatorname{sgn}(xy) = \operatorname{sgn}(x) \cdot \operatorname{sgn}(y)$
4. $\operatorname{sgn}\left(\dfrac{y}{x}\right) = \operatorname{sgn}(xy)$ if $x \neq 0$

The proofs are straightforward so we leave them to the reader. As a hint, the second property is proven by casework, and the third and fourth properties follow from it.

6.2 Rounding

When working with a non-integer real number, we sometimes want to work with a nearby integer instead. This leads to several notions.

Definition 6.9. The **floor** function, **ceiling** function, and **fractional part** of a real number are respectively defined as follows:

$$\lfloor x \rfloor = \max\{m \in \mathbb{Z} : m \leq x\},$$
$$\lceil x \rceil = \min\{n \in \mathbb{Z} : n \geq x\},$$
$$\{x\} = x - \lfloor x \rfloor.$$

For those familiar with modular arithmetic from Volume 3, it will be evident that taking the fractional part of a real number is akin to a "reduction modulo 1" function for real numbers.

Example. Let's compute the floor, ceiling, and fractional part of -2.6. It is clear that $\lfloor -2.6 \rfloor = -3$ and $\lceil -2.6 \rceil = -2$. Perhaps counterintuitively,

$$\{-2.6\} = -2.6 - \lfloor -2.6 \rfloor = -2.6 - (-3) = -2.6 + 3 = 0.4,$$

which is neither -0.6 nor 0.6.

Problem 6.10. Prove that, for any integer n,

$$\left\lfloor \frac{n}{2} \right\rfloor + \left\lceil \frac{n}{2} \right\rceil = n \text{ and } \left\lfloor \frac{n}{2} \right\rfloor \cdot \left\lceil \frac{n}{2} \right\rceil = \left\lfloor \frac{n^2}{4} \right\rfloor.$$

Lemma 6.11. For each real number x, there exists exactly one integer in each of the intervals $[x, x+1)$ and $(x-1, x]$.

Proof. We will show the proof for $[x, x+1)$ and leave the proof for $(x-1, x]$ to the reader, since the latter is analogous to the former. Suppose, for contradiction, that there is no integer in the interval $[x, x+1)$. Let n be the greatest integer that is less than x. Then $n < x$ and, since there is no integer in $[x, x+1)$, the next integer is $n+1 \geq x+1$. But this means $n \geq x$, which contradicts $n < x$. So there must exist an integer in $[x, x+1)$. Now suppose there are two integers n and m in $[x, x+1)$. Then we have the inequalities

$$x \leq n < x+1,$$
$$x \leq m < x+1.$$

The negative of the latter is $-x-1 < -m \leq -x$, and adding this to the first inequality yields

$$-1 < n - m < 1.$$

But $n-m$ is an integer, and the only integer that is strictly between -1 and 1 is 0. Thus $n = m$, and so there is a unique integer in $[x, x+1)$. □

Theorem 6.12. Let m and n be integers and x be a real number. Then the following two sequences of equivalences hold:

$$\lfloor x \rfloor = m \iff x - 1 < m \leq x \iff m \leq x < m+1,$$
$$\lceil x \rceil = n \iff x \leq n < x+1 \iff n-1 < x \leq n.$$

Proof. Combining the original definitions of the floor and ceiling functions with Lemma 6.11 allows us to redefine $\lfloor x \rfloor$ as the unique integer in the interval $(x-1, x]$ and $\lceil x \rceil$ as the unique integer in the interval $[x, x+1)$. The first equivalence of each sequence follows from this, and the second equivalence of each sequence is a simple matter of rewriting the inequalities. □

Corollary 6.13. Using Theorem 6.12, we can present the floor and ceiling functions in infinitely many pieces:

$$\lfloor x \rfloor = \begin{cases} \vdots & \\ -2 & \text{if } -2 \leq x < -1 \\ -1 & \text{if } -1 \leq x < 0 \\ 0 & \text{if } 0 \leq x < 1 \\ 1 & \text{if } 1 \leq x < 2 \\ 2 & \text{if } 2 \leq x < 3 \\ \vdots & \end{cases}, \lceil x \rceil = \begin{cases} \vdots & \\ -2 & \text{if } -3 < x \leq -2 \\ -1 & \text{if } -2 < x \leq -1 \\ 0 & \text{if } -1 < x \leq 0 \\ 1 & \text{if } 0 < x \leq 1 \\ 2 & \text{if } 1 < x \leq 2 \\ \vdots & \end{cases}.$$

Corollary 6.14. For any real number x, it holds that

$$0 \leq \{x\} < 1.$$

Proof. This follows from combining the definition $\{x\} = x - \lfloor x \rfloor$ with the fact that, if $\lfloor x \rfloor = m$, then $m \leq x < m + 1$ or $0 \leq x - m < 1$ (Theorem 6.12). $\square$

Problem 6.15. Prove that

$$\lfloor x \rfloor = -\lceil -x \rceil$$

for all real x. In terms of graphing on the Cartesian plane, this means that applying a reflection across the x-axis and also across the y-axis allows us to switch between the floor function and ceiling function.

Theorem 6.16. If x is a real number and p is an integer, then:

$$\begin{aligned} x < p &\iff \lfloor x \rfloor < p, \\ p \leq x &\iff p \leq \lfloor x \rfloor, \\ p < x &\iff p < \lceil x \rceil, \\ x \leq p &\iff \lceil x \rceil \leq p. \end{aligned}$$

These equivalences among inequalities are interesting and applicable, as they allow us to switch a variable with its ceiling or floor when being compared to an integer.

Proof. One direction of each biconditional statement is trivial and the other direction makes the inequality sharper. We will prove the first two equivalences and we leave the remaining two as exercises to the reader, as they require arguments similar to those used to prove the first two.
Since $\lfloor x \rfloor \leq x$, it follows immediately that $x < p$ implies $\lfloor x \rfloor < p$. Conversely, suppose $\lfloor x \rfloor < p$. Since $\lfloor x \rfloor$ and p are both integers, Theorem 5.18 tells us that $\lfloor x \rfloor \leq p - 1$ or $\lfloor x \rfloor + 1 \leq p$. If $\lfloor x \rfloor = m$ then we know that $m \leq x < m + 1 \leq p$. Thus, $x < p$.
For the second equivalence, it is immediately true that $p \leq \lfloor x \rfloor$ implies $p \leq x$. Now suppose $p \leq x$. Then p is an element of the set $\{m \in \mathbb{Z} : m \leq x\}$. By definition, $\lfloor x \rfloor$ is the maximal element of this set, so it must be true that $p \leq \lfloor x \rfloor$, as desired. $\square$

Problem 6.17. Show that the following identities hold for integers $n > 0$ and m:

$$\left\lfloor \frac{m}{n} \right\rfloor = \left\lceil \frac{m - n + 1}{n} \right\rceil \quad \text{and} \quad \left\lceil \frac{m}{n} \right\rceil = \left\lfloor \frac{m + n - 1}{n} \right\rfloor.$$

These allow us to change between floor and ceiling functions for rational inputs.

Theorem 6.18. There is a useful technique of extracting integers out of the floor and ceiling functions and fractional part. If x is a real number and p is an integer, then:

$$\begin{aligned} \lfloor x + p \rfloor &= \lfloor x \rfloor + p, \\ \lceil x + p \rceil &= \lceil x \rceil + p, \\ \{x + p\} &= \{x\}. \end{aligned}$$

Proof. For the floor function, let $\lfloor x + p \rfloor = m$. Then $m \leq x + p < m + 1$ or

$$m - p \leq x < m - p + 1.$$

This means $\lfloor x \rfloor = m - p$, which is equivalent to the desired identity. Using this identity allows us to show that

$$\begin{aligned} \{x + p\} &= x + p - \lfloor x + p \rfloor \\ &= x + p - \lfloor x \rfloor - p \\ &= x - \lfloor x \rfloor \\ &= \{x\}. \end{aligned}$$

The ceiling function's proof is analogous to the proof for the floor function. $\square$

Example 6.19. The usual rounding function takes 6.2 to 6, and 6.7 to 7, and 6.5 to 7. Show that the rounding function may be encapsulated by the definition

$$\mathrm{round}(x) = \left\lfloor x + \frac{1}{2} \right\rfloor.$$

Solution. Indeed, we find that

$$\begin{aligned} \left\lfloor x + \frac{1}{2} \right\rfloor &= \left\lfloor \lfloor x \rfloor + \{x\} + \frac{1}{2} \right\rfloor \\ &= \lfloor x \rfloor + \left\lfloor \{x\} + \frac{1}{2} \right\rfloor \\ &= \begin{cases} \lfloor x \rfloor & \text{if } 0 \le \{x\} < 0.5 \\ \lfloor x \rfloor + 1 & \text{if } 0.5 \le \{x\} < 1 \end{cases}, \end{aligned}$$

which matches the ordinary decision-making process for rounding. □

Theorem 6.20 (Hermite's identity)**.** For all real x and positive integers m,

$$\lfloor x \rfloor + \left\lfloor x + \frac{1}{m} \right\rfloor + \left\lfloor x + \frac{2}{m} \right\rfloor + \cdots + \left\lfloor x + \frac{m-1}{m} \right\rfloor = \lfloor mx \rfloor.$$

In particular, we can take $x = \dfrac{n}{m}$ for an integer n to produce an interesting identity.

Proof. Fix a positive integer m and let

$$f(x) = \lfloor x \rfloor + \left\lfloor x + \frac{1}{m} \right\rfloor + \cdots + \left\lfloor x + \frac{m-2}{m} \right\rfloor + \left\lfloor x + \frac{m-1}{m} \right\rfloor - \lfloor mx \rfloor.$$

We want to show that f is identically 0. Miraculously, we find that f is periodic (periodicity is defined in Definition 8.1) because

$$\begin{aligned} f\left(x + \frac{1}{m}\right) &= \left\lfloor x + \frac{1}{m} \right\rfloor + \left\lfloor x + \frac{2}{m} \right\rfloor + \cdots + \left\lfloor x + \frac{m-1}{m} \right\rfloor + \lfloor x + 1 \rfloor - \lfloor mx + 1 \rfloor \\ &= \left\lfloor x + \frac{1}{m} \right\rfloor + \left\lfloor x + \frac{2}{m} \right\rfloor + \cdots + \left\lfloor x + \frac{m-1}{m} \right\rfloor + \lfloor x \rfloor - \lfloor mx \rfloor \\ &= f(x). \end{aligned}$$

So, it suffices to show that f is identically 0 on the interval $\left[0, \dfrac{1}{m}\right)$. For any x in this interval, it turns out that every term in the expression for f vanishes because of the following reasoning. Since $0 \le x < \dfrac{1}{m}$,

$$0 \le \frac{k}{m} \le x + \frac{k}{m} < \frac{k+1}{m} \le 1.$$

This means $0 \le x + \dfrac{k}{m} < 1$ or $\left\lfloor x + \dfrac{k}{m} \right\rfloor = 0$ for $k = 0, 1, \ldots, m-1$. Moreover, if $0 \le x < \dfrac{1}{m}$ then $0 \le mx < 1$ which tells us that $\lfloor mx \rfloor = 0$. This completes the proof.

Note that the identity

$$\lceil x \rceil + \left\lceil x - \frac{1}{m} \right\rceil + \left\lceil x - \frac{2}{m} \right\rceil + \cdots + \left\lceil x - \frac{m-1}{m} \right\rceil = \lceil mx \rceil$$

for ceiling functions can be derived by applying Hermite's Identity to $-x$ and using the fact that $\lceil x \rceil = -\lfloor -x \rfloor$ (Problem 6.15). We did not include it in the statement of the theorem because it is neither widely cited, nor is it particularly applicable. □

Theorem 6.21. For any real x and y, the following inequalities hold:

$$\lfloor x \rfloor + \lfloor y \rfloor \leq \lfloor x+y \rfloor \leq \lfloor x \rfloor + \lfloor y \rfloor + 1,$$
$$\lceil x \rceil + \lceil y \rceil - 1 \leq \lceil x+y \rceil \leq \lceil x \rceil + \lceil y \rceil.$$

Proof. We will show that the floor function inequalities hold. Note that

$$\lfloor x+y \rfloor = \lfloor \lfloor x \rfloor + \{x\} + \lfloor y \rfloor + \{y\} \rfloor = \lfloor x \rfloor + \lfloor y \rfloor + \lfloor \{x\} + \{y\} \rfloor.$$

So it suffices to prove

$$0 \leq \lfloor \{x\} + \{y\} \rfloor \leq 1.$$

Since $\lfloor \{x\} + \{y\} \rfloor$ is an integer, it is equivalent to show that $\lfloor \{x\} + \{y\} \rfloor$ is either 0 or 1. Since $0 \leq \{x\} < 1$ and $0 \leq \{y\} < 1$, adding them yields $0 \leq \{x\} + \{y\} < 2$. Indeed, the possibilities then are

$$\lfloor \{x\} + \{y\} \rfloor = \begin{cases} 0 & \text{if } 0 \leq \{x\} + \{y\} < 1, \\ 1 & \text{if } 1 \leq \{x\} + \{y\} < 2. \end{cases}$$

The ceiling function inequalities follow from applying the floor ones to $-x$ and $-y$ because

$$\lceil x \rceil = -\lfloor -x \rfloor,$$

by Problem 6.15. □

Corollary 6.22. The floor and ceiling functions have the non-decreasing property, meaning if x and y are real numbers such that $x < y$ then

$$\lfloor x \rfloor \leq \lfloor y \rfloor \text{ and } \lceil x \rceil \leq \lceil y \rceil.$$

Proof. Let $k > 0$ be defined as $y = x + k$. Since $k > 0$ means $\lfloor k \rfloor \geq 0$, the floor function inequality in Theorem 6.21 tells us that

$$\lfloor x \rfloor \leq \lfloor x \rfloor + \lfloor k \rfloor \leq \lfloor x+k \rfloor = \lfloor y \rfloor.$$

Similarly, since $k > 0$ means $\lceil k \rceil \geq 1$, the ceiling function inequality in Theorem 6.21 tells us that

$$\lceil x \rceil \leq \lceil x \rceil + \lceil k \rceil - 1 \leq \lceil x+k \rceil = \lceil y \rceil.$$

□

We end the section with a general theorem. It has several ramifications, as we will see in Corollary 6.24. Notice that the floor and ceiling functions are idempotent, meaning

$$\lfloor \lfloor x \rfloor \rfloor = \lfloor x \rfloor \text{ and } \lceil \lceil x \rceil \rceil = \lceil x \rceil.$$

We can ask whether we can insert a function in between the two applications of the floor or ceiling function and still retain equality.

Theorem 6.23 (McEliece's theorem)**.** Let f be a continuous and monotonically increasing function, defined on some interval $I \subseteq \mathbb{R}$, with the property that

$$f(x) \in \mathbb{Z} \implies x \in \mathbb{Z}.$$

Then, for all $x \in I$,

$$\lfloor f(x) \rfloor = \lfloor f(\lfloor x \rfloor) \rfloor \text{ and } \lceil f(x) \rceil = \lceil f(\lceil x \rceil) \rceil.$$

Of course, each identity holds only if the expression on either side of it is defined, as x being in I does not necessarily mean that $\lfloor x \rfloor$ and $\lceil x \rceil$ are in I.

Proof. We will prove the theorem for floor functions and leave the analogous argument for ceiling functions as an exercise to the reader. The reader may also find the proof for ceiling functions in [6].
So, we want to prove that $\lfloor f(x)\rfloor = \lfloor f(\lfloor x\rfloor)\rfloor$. Recall that $\lfloor x\rfloor \leq x$. If $\lfloor x\rfloor = x$ then the identity immediately holds. Now suppose $\lfloor x\rfloor < x$, implying x is not an integer. By the fact that f is monotonically increasing,

$$f(\lfloor x\rfloor) < f(x).$$

Since the floor function is non-decreasing, we get

$$\lfloor f(\lfloor x\rfloor)\rfloor \leq \lfloor f(x)\rfloor.$$

If equality holds in this inequality, then we are done. So suppose for contradiction that

$$\lfloor f(\lfloor x\rfloor)\rfloor < \lfloor f(x)\rfloor.$$

Then $\lfloor f(\lfloor x\rfloor)\rfloor < \lfloor f(x)\rfloor \leq f(x)$. Now there are two possibilities, both of which will lead to the same conclusion. One possibility is that $\lfloor f(x)\rfloor = f(x)$. The other possibility is that $\lfloor f(\lfloor x\rfloor)\rfloor < \lfloor f(x)\rfloor < f(x)$. In the latter case, since $\lfloor f(x)\rfloor$ is an integer, we can use Theorem 6.16 to take off the floor function on the left side and get

$$f(\lfloor x\rfloor) < \lfloor f(x)\rfloor < f(x).$$

In this case, we can invoke continuity of f to use the intermediate value theorem from calculus, which tells us here that there exists $y \in (\lfloor x\rfloor, x)$ such that $f(y) = \lfloor f(x)\rfloor$.
In either case, $\lfloor f(x)\rfloor = f(z)$ for some z. Since $f(z) = \lfloor f(x)\rfloor$ is an integer, z must be an integer by the special property of f. But $\lfloor x\rfloor < z \leq x$, so z cannot be an integer since x is assumed to not be an integer and $\lfloor x\rfloor$ is the maximum integer that is less than or equal to x. So $\lfloor f(\lfloor x\rfloor)\rfloor < \lfloor f(x)\rfloor$ leads to a contradiction, forcing it to be the case that $\lfloor f(\lfloor x\rfloor)\rfloor = \lfloor f(x)\rfloor$. □

Corollary 6.24. There are several immediate examples of continuous and monotonically increasing functions f that are defined on an interval $I \subseteq \mathbb{R}$ and satisfy the property

$$f(x) \in \mathbb{Z} \implies x \in \mathbb{Z}:$$

- $f(x) = \sqrt[n]{x}$ for integers $n \geq 1$, where the domain is $[0, \infty)$ for even n and $\mathbb{R}$ for odd n
- $f(x) = \log_b x$ for integer bases $b \geq 2$, where the domain is $[1, \infty)$
- $f(x) = \dfrac{x+m}{n}$ for integers $n > 0$ and m, where the domain is $\mathbb{R}$
  In particular, if we choose $m = 0$ and a sequence of positive integers $n_1, n_2, \ldots, n_k$ and replace x with $\dfrac{x}{n_1}$, then repeated application of McEliece's theorem yields the nested division formulas

$$\lfloor \ldots \lfloor \lfloor x/n_1\rfloor /n_2\rfloor \ldots /n_k\rfloor = \left\lfloor \frac{x}{n_1 n_2 \cdots n_k}\right\rfloor,$$
$$\lceil \ldots \lceil \lceil x/n_1\rceil /n_2\rceil \ldots /n_k\rceil = \left\lceil \frac{x}{n_1 n_2 \cdots n_k}\right\rceil.$$

Chapter 7

Closed Forms

"Most of the time, I know what to do. I don't have to figure it out. I don't have to sit there, calculate for forty-five minutes, an hour, to know what is the right move. I just, usually I can just feel it immediately... I have to, you know, verify my opinion, see that I haven't missed anything. But a lot of the time, it's fairly useless because I know what I'm going to do, and then I sit there for a long time, and I do what I immediately wanted to do."

– *Magnus Carlsen, 60 Minutes Overtime*

"Do not be afraid to make mistakes. A mistake for a chess player is fatal; for a mathematician it is par for the course. What you should be terrified of is a blank sheet in front of you after having thought about a problem for a little while."

– *Béla Bollob as*

Having studied sequences, sums, and products in general in Chapter 4, we will now take a look at particular instances of them that are amenable to expression in closed form. First we will look at arithmetic and geometric sequences and series. Then we will move on to studying a curious technique called telescoping that is surprisingly applicable.

7.1 Arithmetic and Geometric

Given a finite series or product, it is desirable to find a "closed formula" for it, and similarly for the partial sums or products of an infinite series or product. How a closed formula can be found, if at all possible, depends on the definition of the indexed object. We will find formulas for two of the most common sequences and series, namely arithmetic and geometric sequence and series. There is no universally accepted definition of a closed formula, but we can encapsulate the sentiment informally as follows.

Definition 7.1. A formula is said to be **closed** or be in **closed form** if the number of operations in it is constant or has a constant upper bound across all values of all variables.

Example. The formula $\underbrace{p + p + \cdots + p}_{n \text{ copies of } p}$ is not closed because there are $n-1$ operations, which grows without an upper bound as n goes to infinity. On the other hand, pn is considered to be a closed expression for the same quantity. What exactly is meant by "the number of operations" is ambiguous in Definition 7.1. For example, multiplying two very large integers takes a computer algorithm more steps than multiplying two small integers, but both computations could be considered to be one multiplication operation. Moreover, many "closed" formulas in higher mathematics involve integrals or use functions, such as the Gamma function Γ or the Riemann zeta function ζ, even though it might be difficult to evaluate the outputs of these functions at given inputs.

Definition 7.2. The **arithmetic sequence** with **initial term** a and **common difference** d is

$$(a + (n-1)d)_{n=1}^{\infty} = (a, a+d, a+2d, \ldots).$$

For each positive integer n, the sum of the first n terms of an arithmetic sequence is called an **arithmetic series**.

Theorem 7.3. If the initial term of an arithmetic sequence $(a_n)_{n=1}^{\infty}$ is a and common difference is d, then the resulting arithmetic series with n terms evaluates to

$$a_1 + a_2 + \cdots + a_n = \left(\frac{a + a_n}{2}\right) \cdot n = \frac{(2a + (n-1)d)n}{2}$$

for each positive integer n. The first formula is memorable, whereas the second lacks symmetry.

Proof. The sum can be written as

$$\sum_{k=1}^{n} (a + (k-1)d) = an + d \cdot \sum_{k=1}^{n} (k-1).$$

So the computation boils down to finding the sum

$$1 + 2 + \cdots + (n-1).$$

The first idea is one of which the reader should take careful note because it can be useful elsewhere: set the desired quantity equal to a variable. The second idea is to "reflect" the sum by writing it in reverse order. Let

$$S = \sum_{k=1}^{n-1} k = \sum_{k=1}^{n-1} (n-k).$$

Adding the two sums yields

$$2S = \sum_{k=1}^{n-1} k + \sum_{k=1}^{n-1} (n-k) = \sum_{k=1}^{n-1} n = (n-1)n.$$

So, the sum of the first $n-1$ positive integers is $S = \dfrac{(n-1)n}{2}$. This allows to conclude that

$$\sum_{k=1}^{n} (a + (k-1)d) = an + d \cdot \frac{(n-1)n}{2} = \frac{(2a + (n-1)d)n}{2} = \frac{(a + a_n)n}{2}.$$

According to legend, Gauss was told by his teacher to evaluate the sum $1+2+3+\cdots+100$, and the teacher was astonished when young Gauss quickly produced the answer by using this reflection method. For this reason, the method is sometimes called **Gauss's trick**. □

Definition 7.4. For each positive integer n, the nth **triangular number** is

$$1 + 2 + \cdots + n = \frac{n(n+1)}{2}.$$

The name is derived from the fact that $1 + 2 + \cdots + n$ dots can be stacked in a triangular array as shown below.

Problem 7.5. For each positive integer n, find the sum of the first n odd positive integers

$$1 + 3 + 5 + \cdots + (2n-1).$$

Definition 7.6. The **geometric sequence** with **initial term** b and **common ratio** r is

$$(br^{n-1})_{n=1}^{\infty} = (b, br, br^2, \ldots).$$

For each positive integer n, the sum of the first n terms of a geometric sequence is called a **geometric series**. An **infinite geometric series** is the sum of *all* terms in a geometric sequence; this does not always converge to a number.

Theorem 7.7. If the initial term of a geometric sequence

$$(b_n)_{n=1}^{\infty} = (b_1, b_2, b_3, \ldots)$$

is b and common ratio is r, then the resulting geometric series with n terms evaluates to

$$b_1 + b_2 + \cdots + b_n = \begin{cases} bn & \text{if } r = 1 \\ \dfrac{b(r^n - 1)}{r - 1} & \text{if } r \neq 1 \end{cases}$$

for each positive integer n. If $b = 0$ then the infinite geometric series is 0, so suppose $b \neq 0$. If $|r| < 1$, then this infinite sum converges to

$$\sum_{k=0}^{\infty} br^k = \frac{b}{1 - r}.$$

However, the infinite geometric series does not converge to a number if $|r| \geq 1$.

Proof. First, we tackle the finite case. The formula for $r = 1$ is obviously true, so we assume that $r \neq 1$. As with an arithmetic series, we set the desired quantity equal to a variable. We let

$$S = b + br + br^2 + \cdots + br^{n-1}.$$

Then we create another sum that has many of the same terms, so that we can take the difference of the two. Multiplying the first equation by r yields

$$rS = br + br^2 + br^3 + \cdots + br^n.$$

Subtracting the first equation from the second gives

$$\begin{aligned}(r - 1)S &= rS - S \\ &= (br + br^2 + br^3 + \cdots + br^n) - (b + br + br^2 + \cdots + br^{n-1}) \\ &= br^n - b = b(r^n - 1).\end{aligned}$$

By isolating S, we get the desired formula

$$S = \frac{b(r^n - 1)}{r - 1}.$$

Now, we will work on the infinite case with $b \neq 0$. The following are the cases of $|r| \geq 1$.

- If $r = 1$, then the series has partial sums bn. If $b > 0$, then bn grows without an upper bound as n goes to infinity; if $b < 0$, then bn goes to $-\infty$ as n goes to infinity.
- If $r = -1$, then the partial sums oscillate between b and $-b$, which also does not converge.
- If $r > 1$, then the partial sum $\dfrac{b(r^n - 1)}{r - 1}$ has a constant denominator and a numerator that grows arbitrarily large as n goes to infinity.
- If $r < -1$, then the partial sum $\dfrac{b(r^n - 1)}{r - 1}$ again has a constant denominator, and a numerator that grows arbitrarily large for even integers n as n goes to infinity.

So there is no convergence for $|r| \geq 1$. Finally, we consider the "Goldilocks zone," $|r| < 1$, which is equivalent to $-1 < r < 1$. Here, we rewrite the partial sum as

$$\frac{b(r^n - 1)}{r - 1} = \frac{br^n}{r - 1} + \frac{b}{1 - r}.$$

As n goes to infinity, the first term goes to 0 because its numerator gets arbitrarily close to 0 while the denominator is constant; the second term is constant. This means the partial sums converge to $\dfrac{b}{1 - r}$. Making these arguments more precise would require clearly defining convergence and divergence, which would take us to the realm of limits from calculus. □

Problem 7.8. Let p and $r \neq 1$ be real constants and n be a positive integer. Evaluate

$$\underbrace{(\cdots((pr+p)r+p)r\cdots+p)r+p}_{\text{number of multiplications by } r \text{ is } n-1}.$$

Example 7.9. Prove the following facts about arithmetic and geometric sequences.

1. Let $(a_n)_{n=1}^{\infty}$ be an arithmetic sequence. For any distinct indices i and j, the common difference is $\frac{a_i - a_j}{i-j}$. Also, for any index $n > 1$, a_n is the arithmetic mean of its neighbours, meaning

$$a_n = \frac{a_{n-1} + a_{n+1}}{2}.$$

2. Let $(b_n)_{n=1}^{\infty}$ be a geometric sequence with non-zero initial term and common ratio r. For any distinct indices i and j, $\frac{b_i}{b_j} = r^{i-j}$; in some cases, such as if r is known to be positive or $i-j$ is an odd integer, it is possible to isolate r using a radical. Also, for any index $n > 1$, if b_n is known to be positive (for example, if the initial term and common ratio are both positive), then b_n is the geometric mean of its neighbours, meaning

$$b_n = \sqrt{b_{n-1}b_{n+1}}.$$

Solution. These are all matters of straightforward computation by unwrapping definitions.

1. Let the initial term of the arithmetic sequence be a and the common difference be d. Then

$$\frac{a_i - a_j}{i-j} = \frac{(a+(i-1)d) - (a+(j-1)d)}{i-j} = d.$$

For the second result,

$$\begin{aligned}\frac{a_{n-1} + a_{n+1}}{2} &= \frac{(a+(n-1-1)d) + (a+(n+1-1)d)}{2} \\ &= a + (n-1)d = a_n.\end{aligned}$$

2. Let the initial term of the geometric sequence be $b \neq 0$ and the common ratio be $r \neq 0$. Then

$$\frac{b_i}{b_j} = \frac{br^{i-1}}{br^{j-1}} = r^{(i-1)-(j-1)} = r^{i-j}.$$

For the second result,

$$b_{n-1}b_{n+1} = (br^{n-1-1})(br^{n+1-1}) = (br^{n-1})^2 = b_n^2.$$

If we know that b_n is positive, then we can take the positive square root of both sides to get

$$b_n = \sqrt{b_{n-1}b_{n+1}}.$$

Otherwise, there remains the possibility that $b_n = -\sqrt{b_{n-1}b_{n+1}}$.

□

Definition 7.10. An **arithmetico-geometric sequence** results from multiplying each term of an arithmetic sequence with the corresponding term of a geometric sequence. So if the arithmetic sequence has initial term a and common difference d, and if the geometric sequence has initial term b and common ratio r, then the arithmetico-geometric sequence is

$$((a+(n-1)d) \cdot br^{n-1})_{n=1}^{\infty}.$$

An **arithmetico-geometric series** is defined as one would expect.

Theorem 7.11. The arithmetico-geometric series of $((a + (n-1)d) \cdot br^{n-1})_{n=1}^{\infty}$ up to n terms can be evaluated as

$$\sum_{k=1}^{n} (a + (k-1)d) \cdot br^{k-1} = \frac{ab - (a+nd)br^n}{1-r} + \frac{dbr(1-r^n)}{(1-r)^2},$$

assuming $r \neq 1$. If $r = 1$, then this is simply an arithmetic sequence.

Proof. We will use the same technique that is used to find the sum of a geometric series. Suppose $r \neq 1$. We want to find a closed form for

$$S = ab + (a+d)br + (a+2d)br^2 + \cdots + [a + (n-1)d] \cdot br^{n-1}.$$

Multiplying both sides of this equation by the common ratio r of the associated geometric sequence yields

$$Sr = abr + (a+d)br^2 + \cdots + [a + (n-2)d] \cdot br^{n-1} + [a + (n-1)d] \cdot br^n.$$

Subtracting the latter equation from the former yields

$$\begin{aligned} S(1-r) &= S - Sr \\ &= ab + (dbr + dbr^2 + \cdots + dbr^{n-1}) - [a + (n-1)d] \cdot br^n \\ &= ab + (dbr + dbr^2 + \cdots + dbr^{n-1} + dbr^n) - (abr^n + ndbr^n) \\ &= ab - (a+nd)br^n + \frac{dbr(1-r^n)}{1-r}, \end{aligned}$$

where we used the formula for a geometric series in the last step. Dividing both sides by $1-r$ gives the final formula. It is not necessary to memorize this complicated formula, considering the relatively low frequency with which arithmetico-geometric series appear in practice. We have derived it only to showcase the technique, which can be replicated easily. □

Theorem 7.12. If $|r| < 1$, then

$$\sum_{k=1}^{\infty} (a + (k-1)d) \cdot br^{k-1} = \frac{ab}{1-r} + \frac{dbr}{(1-r)^2}.$$

We could classify all cases of convergence and divergence, like with geometric series, but the casework is not worth the effort in this context.

Proof. The infinite sum is

$$\begin{aligned} \sum_{k=1}^{\infty} (a + (k-1)d) \cdot br^{k-1} &= \lim_{n\to\infty} \sum_{k=1}^{n} (a + (k-1)d) \cdot br^{k-1} \\ &= \lim_{n\to\infty} \left[\frac{ab - (a+nd)br^n}{1-r} + \frac{dbr(1-r^n)}{(1-r)^2}\right] \\ &= \frac{ab}{1-r} + \frac{dbr}{(1-r)^2} - \lim_{n\to\infty} \frac{(a+nd)br^n}{1-r} - \lim_{n\to\infty} \frac{dbr^{n+1}}{(1-r)^2}. \end{aligned}$$

In the second limit, the denominator is constant and the numerator goes to 0 as n goes to infinity. In the first limit, the denominator is again 0, but the numerator is a bit trickier. Without getting into the formal definition of limits, the crux of the matter is that, while $a + nd$ goes to infinity linearly as n goes to infinity, br^n goes to 0 at an exponential rate as n goes to infinity. So the former is overtaken by the latter and the whole term disappears. Thus, we are left with only the two terms outside the limits. □

7.2 Telescoping

Definition 7.13. A series is said to **telescope** if it can be written in a way that many terms cancel each other out. Just like sums, it is possible for products to telescope. The name is derived from how it is possible to collapse some telescopes so that intermediate components disappear from external view.

Example. The product

$$\frac{1}{\cancel{2}} \cdot \frac{\cancel{2}}{\cancel{3}} \cdot \frac{\cancel{3}}{\cancel{4}} \cdots \frac{\cancel{n-1}}{\cancel{n}} \cdot \frac{\cancel{n}}{n+1} = \frac{1}{n+1}$$

telescopes because each denominator cancels with the subsequent numerator, leaving us with only the leftmost numerator and rightmost denominator at the end.

Problem 7.14. Prove that, for any positive integer n,

$$1 \cdot 1! + 2 \cdot 2! + 3 \cdot 3! + \cdots + n \cdot n! = (n+1)! - 1.$$

We will ask the reader for a combinatorial interpretation of this identity in Volume 2.

A telescoping series sometimes goes hand-in-hand with a technique called partial fraction decomposition, which will be used much later to prove Binet's formula for the Fibonacci numbers in Volume 2.

Definition 7.15. Without defining **partial fraction decomposition** in great detail, the general idea is that, if we have an expression of the form $\dfrac{f(x)}{g_1(x)g_2(x)\cdots g_n(x)}$ for functions f and $g_1, g_2, \ldots, g_n$, then it could be useful to decompose it as

$$\frac{f(x)}{g_1(x)g_2(x)\cdots g_n(x)} = \frac{f_1(x)}{g_1(x)} + \frac{f_2(x)}{g_2(x)} + \cdots + \frac{f_n(x)}{g_n(x)}$$

for suitable functions $f_1, f_2, \ldots, f_n$.

Of course, the question is how the f_i can be found. There exists a general theorem for rational functions (see Definition 11.35) that is useful in the integration of such functions in calculus. For our purposes, it will suffice to deal only with distinct linear functions $g_i(x) = a_i x + b_i$ and the constant function $f(x) = 1$. In such cases, we will commit three mathematical sins to find the f_i:

- We will assume that the f_i exist.
- We will assume that the f_i are equal to constants c_i.
- We will clear the denominators and solve for the f_i by substituting in "forbidden" values for x, such as $-\dfrac{b_i}{a_i}$, that would have originally caused

  $$g_1(x)g_2(x)\cdots g_n(x) = 0,$$

  which might as well be fraternizing with division by 0.

Why is all this acceptable? Technically, it has to do with the fundamental theorem of algebra (Theorem 11.28), for which we are not yet ready. However, as long as we find f_i that work in the end, the process can be swept under the rug. Let us see a classic example of a series that can be made to telescope using partial fraction decomposition.

Example 7.16. If n is a positive integer, then find a closed expression for

$$\frac{1}{1 \cdot 2} + \frac{1}{2 \cdot 3} + \frac{1}{3 \cdot 4} + \cdots + \frac{1}{n(n+1)}$$

in terms of n.

Solution. Each term is of the form $\frac{1}{k(k+1)}$. We start by assuming that there exist constants A and B such that

$$\frac{A}{k}+\frac{B}{k+1}=\frac{1}{k(k+1)}.$$

Clearing the denominators yields

$$A(k+1)+Bk=1.$$

If we want A to disappear, then we can substitute $k=-1$, which gives $B=-1$. If we want B to disappear, then we can substitute $k=0$, which gives $A=1$. Our hope that it is in fact true that

$$\frac{1}{k}-\frac{1}{k+1}=\frac{1}{k(k+1)},$$

which we can verify by adding the fractions on the left side. This means the series in the problem becomes

$$\left(\frac{1}{1}-\cancel{\frac{1}{2}}\right)+\left(\cancel{\frac{1}{2}}-\cancel{\frac{1}{3}}\right)+\left(\cancel{\frac{1}{3}}-\cancel{\frac{1}{4}}\right)+\cdots+\left(\cancel{\frac{1}{n}}-\frac{1}{n+1}\right).$$

Many consecutive terms cancel each other out, leaving only the terms on the far left and far right. This yields a final expression of $1-\frac{1}{n+1}=\frac{n}{n+1}$. □

Problem 7.17. Find the sum of the infinite series

$$\frac{1}{1\cdot 2\cdot 3}+\frac{1}{2\cdot 3\cdot 4}+\frac{1}{3\cdot 4\cdot 5}+\cdots$$

by finding the partial sum of the first n terms and then observing what happens as n goes to infinity.

Next, we (re)turn to the problem of finding a closed form for the sum of the pth powers of the first n positive integers:

$$1^p+2^p+\cdots+n^p.$$

Using Gauss's trick, we have already evaluated the sum for $p=1$ as $\frac{n(n+1)}{2}$ (Theorem 7.3). Now, we consider $p=2$ and $p=3$. For $p=2$, we can use telescoping as follows.

Theorem 7.18. For each positive integer n,

$$1^2+2^2+\cdots+n^2=\frac{n(n+1)(2n+1)}{6}.$$

Proof. The motivation is to build a formula for this expression of squares through the lower instance of Gauss's formula. This will be possible by adding the instances of

$$(k+1)^3-k^3=3k^2+3k+1.$$

We evaluate the sum of this quantity from $k=0$ to $k=n$, in two different ways. On the left side, we get a telescoping series

$$\sum_{k=0}^{n}[(k+1)^3-k^3]=(n+1)^3.$$

Using the right side, we get

$$\begin{aligned}\sum_{k=0}^{n}(3k^2+3k+1)&=3\cdot\sum_{k=0}^{n}k^2+3\cdot\sum_{k=0}^{n}k+\sum_{k=0}^{n}1\\&=3\cdot\sum_{k=1}^{n}k^2+\frac{3n(n+1)}{2}+(n+1).\end{aligned}$$

Equating the two sums allows us to perform the isolation

$$\sum_{k=1}^{n} k^2 = \frac{1}{3} \cdot \left[(n+1)^3 - \frac{3n(n+1)}{2} - (n+1) \right] = \frac{n(n+1)(2n+1)}{6}.$$

We leave it to the reader to perform the algebraic manipulations needed to achieve the neat final expression shown.
This overall method can be generalized to recursively find formulas for all positive integers p using the binomial theorem. We will do so in Volume 2. □

Problem 7.19. Miraculously, it holds that

$$1^3 + 2^3 + \cdots + n^3 = (1 + 2 + \cdots + n)^3 = \left[\frac{n(n+1)}{2} \right]^2.$$

We invite the reader to prove this by induction and to take note that it is often possible to prove a closed formula by induction if one knows the destination formula. The reader might be interested in knowing that the author of this book proved the following proposition by an induction argument in a paper with Dr. Edward Barbeau [4]:

If $(a_1, a_2, \ldots, a_n)$ is a strictly increasing tuple of n positive integers, then

$$a_1^3 + a_2^3 + \cdots + a_n^3 \geq (a_1 + a_2 + \cdots + a_n)^2,$$

with equality holding if and only if $(a_1, a_2, \ldots, a_n) = (1, 2, \ldots, n)$.

Example 7.20. Show that, if x is a real number such that $|x| < 1$ then

$$\prod_{k=0}^{\infty} (1 + x^{2^k}) = \sum_{k=0}^{\infty} x^k.$$

Solution. By repeated applications of the difference of squares factorization, we can prove by induction on positive integers n that the nth partial product is equal to

$$\begin{aligned}
\prod_{k=0}^{n} (1 + x^{2^k}) &= (1+x)(1+x^2)(1+x^4)\cdots(1+x^{2^n}) \\
&= \frac{(1-x)(1+x)(1+x^2)(1+x^4)\cdots(1+x^{2^n})}{1-x} \\
&= \frac{1 - x^{2^{n+1}}}{1-x} = \frac{1}{1-x} - \frac{x^{2^{n+1}}}{1-x}.
\end{aligned}$$

As $n \to \infty$, the second term goes to 0 because the numerator $x^{2^{n+1}}$ gets arbitrarily close to 0, while the denominator $1 - x$ remains constant. So

$$\prod_{k=0}^{\infty} (1 + x^{2^k}) = \lim_{n\to\infty} \prod_{k=0}^{n} (1 + x^{2^k}) = \frac{1}{1-x}.$$

Since $|x| < 1$, this is the formula for an infinite geometric whose initial term is 1 and common ratio is x. After you read about generating functions in Volume 2, consider why the result here implies the uniqueness of binary representations of non-negative integers.
This was not an example of a telescoping product *per se*, but the way the terms collapsed into each other in the numerator of the partial product is reminiscent of telescoping. □

Problem 7.21. Evaluate the infinite product $\prod_{k=2}^{\infty} \frac{k^3 - 1}{k^3 + 1}$. You should evaluate the finite version up to n multiplicands first and then take the limit as $n \to \infty$.

Chapter 8

Trigonometric Functions

"Young man, in mathematics you don't understand things. You just get used to them."

– *John von Neumann*

We will look at the six trigonometric functions here in terms of unit circle geometry. The trigonometric functions are ubiquitous in mathematics. We will start with a general study of periodic functions, and then see specific properties of the the main trigonometric functions. Afterwards, we will take an exhaustive look at trigonometric identities.

8.1 Periodic Functions

We will begin with some general facts about periodic functions, since it will turn out that trigonometric functions are all periodic.

Definition 8.1. A non-constant function f is said to be **periodic** if its domain is a subset of $\mathbb{R}$ and there exists a positive real constant p such that

$$x \in \text{Dom}(f) \iff x + p \in \text{Dom}(f)$$

for all real x, and

$$f(x+p) = f(x)$$

for all $x \in \text{Dom}(f)$. Such a number p is called a **period** of f. If there is a smallest positive real constant p with this property, then it is called the **minimal period** of f. However, we often simply refer to the minimal period as "the" period, such as in the case of trigonometric functions.

Example. Not every periodic function has a minimal period. For example, the **Dirichlet function** $\delta : \mathbb{R} \to \mathbb{R}$, which is also called the **indicator function** of the rationals, is defined by

$$\delta(x) = \begin{cases} 1 & \text{if } x \in \mathbb{Q} \\ 0 & \text{if } x \notin \mathbb{Q} \end{cases}.$$

For every $r \in \mathbb{Q}$, $x + r$ is rational if x is rational and $x + r$ is irrational if x is irrational. Thus, for every rational r, $\delta(x + r) = \delta(x)$, meaning r is a period. Then, for every period, a smaller period exists, since there exists a sequence of positive rationals $\frac{1}{2}, \frac{1}{3}, \frac{1}{4}, \ldots$ that get arbitrarily close to 0.
On the other hand, it is true that every *continuous* periodic function with domain $\mathbb{R}$ has a minimal period. We will not prove this here since we have not provided a rigorous definition of continuity.

Theorem 8.2. If f is a periodic function with a period p, then $x + pn \in \text{Dom}(f)$ and $f(x + pn) = f(x)$ for all $x \in \text{Dom}(f)$ and all integers n.

Proof. We will prove this by induction on $n \geq 0$ first, and then we will take care of $n \leq 0$ by a separate induction. The result is true for $n = 0$, by the definition of a function. Now suppose there exists an $n \geq 0$ such that $x + pn \in \text{Dom}(f)$ and $f(x + pn) = f(x)$ for all $x \in \text{Dom}(f)$. By the definition of a periodic function, $x + pn + p = x + p(n + 1)$ is in $\text{Dom}(f)$ and

$$f(x + p(n + 1)) = f((x + pn) + p) = f(x + pn) = f(x)$$

for all $x \in \text{Dom}(f)$. This completes the first induction.
Now we will handle $n \leq 0$. Again, the base case is trivial. Now suppose there exists an $n \leq 0$ such that $x + pn \in \text{Dom}(f)$ and $f(x + pn) = f(x)$ for all $x \in \text{Dom}(f)$. By the definition of a periodic function, since $x + pn = x + p(n-1) + p$ is in $\text{Dom}(f)$, so is $x + p(n-1)$, and

$$f(x) = f(x + pn) = f(x + p(n-1) + p) = f(x + p(n-1))$$

for all $x \in \text{Dom}(f)$. This completes the second induction. □

Theorem 8.3. Suppose $f : X \to \mathbb{R}$ is periodic with minimal period p, where X is a subset of $\mathbb{R}$. If $ax + b$ and $cx + d$ are non-constant linear functions with domains $\mathbb{R}$, then

$$k(x) = af(cx + d) + b$$

is periodic with minimal period $\dfrac{p}{|c|}$.

Proof. First, we show that $\dfrac{p}{|c|}$ is a period of k. We will start by proving that

$$x \in \text{Dom}(k) \iff x + \frac{p}{|c|} \in \text{Dom}(k).$$

Since $ax+b$ and $cx+d$ are bijective from $\mathbb{R}$ to $\mathbb{R}$, it is found that $\text{Dom}(k)$ is the preimage of X under $cx+d$, meaning $x \in \text{Dom}(k)$ if and only if $cx+d \in X$. Since f is periodic, $cx+d \in X$ if and only if $cx \pm p + d \in X$, where the $\pm$ sign is taken to be positive if c is positive and negative if c is negative. This is equivalent to

$$c\left(x + \frac{p}{|c|}\right) + d \in X.$$

Again, using the fact that $\text{Dom}(k)$ is the preimage of X under $cx + d$, the last condition is true if and only if $x + \dfrac{p}{|c|} \in \text{Dom}(k)$, as desired. Moreover, $\dfrac{p}{|c|}$ is positive and

$$\begin{aligned} k\left(x + \frac{p}{|c|}\right) &= af\left(c\left(x + \frac{p}{|c|}\right) + d\right) + b \\ &= af(cx \pm p + d) + b \\ &= af(cx + d) + b \\ &= k(x), \end{aligned}$$

where, as before, the $\pm$ sign is positive if c is positive and negative if c is negative.
Now suppose for contradiction that there exists a period q of f that is strictly smaller than $\dfrac{p}{|c|}$. If $x \in \text{Dom}(k)$, then $x + q$ and $x - q$ are also in $\text{Dom}(k)$, and all three lead to the same value in the image of k. Now we have the sequence of implications for all $x \in \text{Dom}(k)$:

$$\begin{aligned} k(x \pm q) &= k(x) \\ af(c(x \pm q) + d) + b &= af(cx + d) + b \\ f(cx + d \pm cq) &= f(cx + d) \\ f(cx + d + |c|q) &= f(cx + d) \end{aligned}$$

For every real y, there exists a real $x = \dfrac{y - d}{c}$ such that $y = cx + d$. So for every real y,

$$f(y + |c|q) = f(y).$$

But $|c|q < p$, which contradicts the minimality of p as a period of f. □

Although trigonometric functions have purely algebraic definitions in terms of infinite series, it will be more illuminating for us to define them geometrically.

Definition 8.4. Recall that the ratio of the circumference of a circle to its diameter is called π, which has the same value for all circles. A **unit circle** has radius 1, so its circumference is

$$\text{circumference} = \pi \cdot \text{diameter} = 2\pi.$$

For this reason, instead of using 360° to measure a full rotation around a circle, we can use 2π **radians** to measure a full rotation, and we can use fractions of 2π radians to measure other sizes of rotation that are less than or greater than a full rotation. With radians, we do not have units, unlike the degrees symbol. Also, if necessary, we can multiply by the conversion factor

$$\frac{180^\circ}{\pi} = 1$$

or its reciprocal to convert between degrees and radians.

We are now ready to define the six trigonometric functions, whose arguments (inputs) will be in radians, unless otherwise specified.

Definition 8.5. Let θ be a real number, which we interpret as an angle in radians. Starting at $(1, 0)$ on the unit circle, we rotate counterclockwise around the origin by a positive angle θ while staying on the unit circle and land on the point (x, y) on the unit circle. If θ is negative, this is a clockwise rotation. Then we can define the following **trigonometric functions**:

1. Sine:

   $\sin\theta = y$

2. Cosine:

   $\cos\theta = x$

3. Tangent:

   $\tan\theta = \dfrac{y}{x} = \dfrac{\sin\theta}{\cos\theta}$

4. Secant:

   $\sec\theta = \dfrac{1}{x} = \dfrac{1}{\cos\theta}$

5. Cosecant:

   $\csc\theta = \dfrac{1}{y} = \dfrac{1}{\sin\theta}$

6. Cotangent:

   $\cot\theta = \dfrac{x}{y} = \dfrac{\cos\theta}{\sin\theta}$

If we draw the line segment from (x, y) to the origin, and draw the perpendicular distance from (x, y) to the x-axis, then the trigonometric functions can be interpreted as ratios between the legs and hypotenuse of the resulting triangle. The caveat to this approach is that we have to allow for negative lengths in the triangles since x or y can be negative, depending on the quadrant of the angle. We also have to allow for degenerate triangles since (x, y) could lie on the x-axis or y-axis.

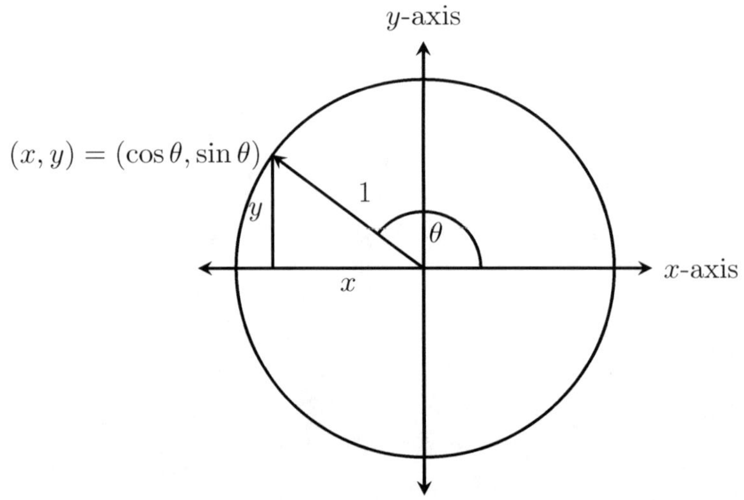

Definition 8.6. If we restrict ourselves to angles in only the first quadrant where the angles are acute and x and y are positive, then we can interpret the trigonometric functions in an equivalent way that has

geometric significance in relation to right triangles. If the angle is θ then there exists a right triangle with angles $\theta, \frac{\pi}{2} - \theta, \frac{\pi}{2}$. Let the side opposite to θ have length o, the adjacent side have length a, and the hypotenuse have length h. This triangle is similar to a triangle with the same angles but with a hypotenuse of length 1 that can be placed inside the first quadrant. The ratios of sides allows us to conclude that:

$$\sin\theta = \frac{o}{h}, \ \cos\theta = \frac{a}{h}, \ \tan\theta = \frac{o}{a}.$$

Similar descriptions hold for the other three (reciprocal) trigonometric functions for acute angles.

Theorem 8.7. See Volume 4 for the fact that $30° - 60° - 90°$ right triangles have sides in the ratio $1 : \sqrt{3} : 2$, and that $45° - 45° - 90°$ right triangles have sides in the ratio $1 : 1 : \sqrt{2}$. These triangles allow us to compute the following special values of the three main trigonometric functions:

Function	$\theta = \frac{\pi}{6}$	$\theta = \frac{\pi}{4}$	$\theta = \frac{\pi}{3}$	$\theta = \frac{\pi}{2}$
sin	$\frac{1}{2}$	$\frac{1}{\sqrt{2}} = \frac{\sqrt{2}}{2}$	$\frac{\sqrt{3}}{2}$	1
cos	$\frac{\sqrt{3}}{2}$	$\frac{1}{\sqrt{2}} = \frac{\sqrt{2}}{2}$	$\frac{1}{2}$	0
tan	$\frac{1}{\sqrt{3}} = \frac{\sqrt{3}}{3}$	1	$\sqrt{3}$	undefined

The trigonometric functions are periodic. This makes sense, since angles that are 2π apart lead to the same coordinates (x, y). However, the period of not every trigonometric function is 2π, as some periods are smaller. Note that, when we discuss trigonometric functions, "the" period refers to the minimal period.

Theorem 8.8. The domains, ranges, periods, and sets of roots of the six trigonometric functions are summarized as follows:

Function	Domain	Range	Period	Roots
sin	$\mathbb{R}$	$[-1, 1]$	2π	$\{\pi n : n \in \mathbb{Z}\}$
cos	$\mathbb{R}$	$[-1, 1]$	2π	$\left\{\frac{\pi}{2} + \pi n : n \in \mathbb{Z}\right\}$
tan	$\mathbb{R} \setminus \left\{\frac{\pi}{2} + \pi n : n \in \mathbb{Z}\right\}$	$\mathbb{R}$	π	$\{\pi n : n \in \mathbb{Z}\}$
sec	$\mathbb{R} \setminus \left\{\frac{\pi}{2} + \pi n : n \in \mathbb{Z}\right\}$	$\mathbb{R} \setminus (-1, 1)$	2π	None
csc	$\mathbb{R} \setminus \{\pi n : n \in \mathbb{Z}\}$	$\mathbb{R} \setminus (-1, 1)$	2π	None
cot	$\mathbb{R} \setminus \{\pi n : n \in \mathbb{Z}\}$	$\mathbb{R}$	π	$\left\{\frac{\pi}{2} + \pi n : n \in \mathbb{Z}\right\}$

Problem 8.9. Use the fact that we can uniquely represent each real number as some $\tan\theta$ within one period of the tangent function to prove that all finite open intervals (a, b) for $a < b$ have the same cardinality as all of $\mathbb{R}$. In other words, find a bijection from (a, b) to $\mathbb{R}$.

Definition 8.10. A **sinusoidal function** is a function that is a result of left-composing or right-composing a non-constant linear function with sine to produce a function of the form

$$f(x) = a\sin(cx + d) + b$$

or similarly with cosine. The **amplitude** of f is half the difference between its largest output and smallest output. Since such a function will inherit the periodicity of sine or cosine by Theorem 8.3, its **period** is defined in the standard way.

Theorem 8.11. The amplitude of the sinusoidal function

$$f(x) = a\sin(cx + d) + b$$

is $|a|$, and its period is $\frac{2\pi}{|c|}$. The same result holds for cosine.

Proof. Since the range of $\sin(cx + d)$ is $[-1, 1]$,

$$\operatorname{Rng}(f) = \begin{cases} [-a + b, a + b] & \text{if } a > 0 \\ [a + b, -a + b] & \text{if } a < 0 \end{cases},$$

both of which can be proven using inequalities. So, if $a > 0$, then the amplitude is half of the difference $(a+b) - (-a+b) = 2a$, and, if $a < 0$, then the amplitude is half of the difference $(-a+b) - (a+b) = -2a$. Either way, the amplitude is $2|a|$. The period follows from the fact that the period of sine is 2π and Theorem 8.3. The logic is identical for cosine. □

Thus far, we have defined the trigonometric functions, computed their values at some common inputs, observed their domains, ranges, and periods, and studied the composition of sine and cosine with linear functions. The next step is to consider whether we can invert trigonometric functions. As a consequence of the periodicity of each trigonometric function f, the preimage of each value in the range of f has infinitely many elements in the domain of f. So, true inverses do not exist. However, we can rectify this situation.

Definition 8.12. Let f be a trigonometric function. Suppose we can restrict f to a domain I such that the $\operatorname{Rng}(f|_I) = \operatorname{Rng}(f)$ and $f|_I$ is injective. Then we can define the function g so that, for each $y \in \operatorname{Rng}(f)$, we choose $g(y)$ to be the unique element x of the preimage $f^{-1}(y)$ that lies in I. We call x the **principal value** of the preimage $f^{-1}(y)$. The set of principal values I of f is the range of g, and by convention I is chosen to an interval containing 0. **Inverse trigonometric functions** are defined as follows:

Function	Definition	Domain	Principal Range
$\arcsin y = x$	$\sin x = y$	$[-1, 1]$	$\left[-\frac{\pi}{2}, \frac{\pi}{2}\right]$
$\arccos y = x$	$\cos x = y$	$[-1, 1]$	$[0, \pi]$
$\arctan y = x$	$\tan x = y$	$\mathbb{R}$	$\left(-\frac{\pi}{2}, \frac{\pi}{2}\right)$

Inverse functions arccot, arcsec, arccsc can also be defined. As they are less common and as there is some disagreement among authors about the choice of principal values for arcsec and arccsc, we have omitted them from the discussion.

8.2 Trigonometric Identities

There are many identities that we can develop involving trigonometric functions. We begin with single-variable identities. In general, we do not allow for division by 0, so the domains of the identities should be restricted as necessary. It is too tedious to always explicitly state the restrictions.

Theorem 8.13 (Pythagorean identity)**.** For each trigonometric function f, let the expression $f^2(\theta)$ refer to the squared expression $(f(\theta))^2$, and not the composition $f(f(\theta))$; this notation is a convention for trigonometric functions. Then

$$\sin^2\theta + \cos^2\theta = 1,$$
$$\tan^2\theta + 1 = \sec^2\theta,$$
$$1 + \cot^2\theta = \csc^2\theta.$$

We will usually not list identities about sec, csc, and cot since they readily reduce to results about sin, cos, and tan, but we have made an exception in this case due to the high frequency with which the Pythagorean identities appear in practice. The functions sec, csc, and cot also sometimes make formulas more confusing than the "simplification" is worth.

Proof. Given an angle θ, let (x, y) be the point $(\cos\theta, \sin\theta)$ on the unit circle, as prescribed by the definition of the trigonometric functions. Then we draw line segment between (x, y) and the origin, and the perpendicular distance from (x, y) to the x-axis. This results in a right triangle with hypotenuse 1 and legs of length $|x|$ and $|y|$. By the Pythagorean theorem,

$$\sin^2\theta + \cos^2\theta = x^2 + y^2 = |x|^2 + |y|^2 = 1.$$

The other two identities follow by dividing the first Pythagorean identity by $\cos^2\theta$ or $\sin^2\theta$. □

Theorem 8.14. The reflection identities are:

	Even-Odd	Complementary Angle	Supplementary Angle
Reflection Across	$y = 0$	$x = y$	$x = 0$
Sine	$\sin(-\theta) = -\sin\theta$	$\sin\left(\frac{\pi}{2} - \theta\right) = \cos\theta$	$\sin(\pi - \theta) = \sin\theta$
Cosine	$\cos(-\theta) = \cos\theta$	$\cos\left(\frac{\pi}{2} - \theta\right) = \sin\theta$	$\cos(\pi - \theta) = -\cos\theta$
Tangent	$\tan(-\theta) = -\tan\theta$	$\tan\left(\frac{\pi}{2} - \theta\right) = \cot\theta$	$\tan(\pi - \theta) = -\tan\theta$

Notice that right-composing a linear function with cosine produces sine and vice versa.

We will not prove these identities here. Although they readily follow from making geometric observations about right triangles in the unit circle, there can be extensive casework on quadrants that is cumbersome enough to not be conducive to our purposes.

Theorem 8.15. This set of identities involve horizontal translations of trigonometric functions by some standard values. The phase shift identities are:

	Shift by $\frac{\pi}{2}$	Shift by π	Shift by 2π
Sine	$\sin\left(\theta \pm \frac{\pi}{2}\right) = \pm\cos\theta$	$\sin(\theta \pm \pi) = -\sin\theta$	$\sin(\theta \pm 2\pi) = \sin\theta$
Cosine	$\cos\left(\theta \pm \frac{\pi}{2}\right) = \mp\sin\theta$	$\cos(\theta \pm \pi) = -\cos\theta$	$\cos(\theta \pm 2\pi) = \cos\theta$
Tangent	$\tan\left(\theta \pm \frac{\pi}{2}\right) = -\cot\theta$	$\tan(\theta \pm \pi) = \tan\theta$	$\tan(\theta \pm 2\pi) = \tan\theta$

Proof. The phase shift identities with shifts by 2π in the right-most column are easy to prove since we know that the periods of sin and cos are 2π, and the period of tan is π. Each tan identity in the bottom row follows from finding the quotient of the analogous identities above it for sine and cosine. The remaining identities follow from applying the reflections identities. For illustrative purposes, here is an example:

$$\begin{aligned}\sin\left(\theta + \frac{\pi}{2}\right) &= \sin\left(\frac{\pi}{2} - (-\theta)\right)\\ &= \cos(-\theta)\\ &= \cos\theta.\end{aligned}$$

The reader should verify the others independently. □

The next step is to jump from single-variable identities to identities in two variables.

Theorem 8.16. It may be desirable to decompose the trigonometric function of a sum or difference of angles into an expression involving the trigonometric functions of the individual angles. This idea is encapsulated in the angle sum and angle difference identities, otherwise known as compound angle identities:

	Angle Sum	Angle Difference
Sine	$\sin(\alpha+\beta) = \sin\alpha\cos\beta + \sin\beta\cos\alpha$	$\sin(\alpha-\beta) = \sin\alpha\cos\beta - \sin\beta\cos\alpha$
Cosine	$\cos(\alpha+\beta) = \cos\alpha\cos\beta - \sin\alpha\sin\beta$	$\cos(\alpha-\beta) = \cos\alpha\cos\beta + \sin\alpha\sin\beta$
Tangent	$\tan(\alpha+\beta) = \dfrac{\tan\alpha+\tan\beta}{1-\tan\alpha\tan\beta}$	$\tan(\alpha-\beta) = \dfrac{\tan\alpha-\tan\beta}{1+\tan\alpha\tan\beta}$

Proof. Taking the first identity

$$\sin(\alpha+\beta) = \sin\alpha\cos\beta + \cos\alpha\sin\beta$$

for granted allows us to derive the rest using it and the reflection identities. □

Corollary 8.17. The double angle and half angle identities are:

	Double Angle	Half Angle
Sine	$\sin(2\theta) = 2\sin\theta\cos\theta$	$\sin^2\frac{\theta}{2} = \frac{1-\cos\theta}{2}$
Cosine	$\cos(2\theta) = \cos^2\theta - \sin^2\theta$ $= 2\cos^2\theta - 1$ $= 1 - 2\sin^2\theta$	$\cos^2\frac{\theta}{2} = \frac{1+\cos\theta}{2}$
Tangent	$\tan(2\theta) = \dfrac{2\tan\theta}{1-\tan^2\theta}$	$\tan\frac{\theta}{2} = \frac{1-\cos\theta}{\sin\theta}$ $= \frac{\sin\theta}{1+\cos\theta}$

Note that taking the plus-minus square root in the half angle formula for sine or cosine leads to two possible isolations. The sign of $\sin\frac{\theta}{2}$ or $\cos\frac{\theta}{2}$ depends on the quadrant in which $\frac{\theta}{2}$ lies.

Proof. The double angle identities follow from substituting in $\theta = \alpha = \beta$ into the identities for $\sin(\alpha+\beta)$, $\cos(\alpha+\beta)$, and $\tan(\alpha+\beta)$. To produce the second and third double angle identities for cosine, we apply the main Pythagorean identity to the first one. The half angle identities for sine and cosine follow from the double angle identities for cosine using a natural angle substitution.
The half angle identities for tangent are more troublesome: taking the quotient of the half angle identities for sine and cosines yields

$$\tan^2\frac{\theta}{2} = \frac{\sin^2\frac{\theta}{2}}{\cos^2\frac{\theta}{2}} = \frac{1-\cos\theta}{1+\cos\theta}.$$

Multiplying the numerator and denominator by $1-\cos\theta$ and applying the Pythagorean identity yields

$$\tan^2\frac{\theta}{2} = \frac{(1-\cos\theta)^2}{1-\cos^2\theta} = \left(\frac{1-\cos\theta}{\sin\theta}\right)^2.$$

Then we notice that $1-\cos\theta$ is always positive, and $\tan\frac{\theta}{2}$ always has the same sign as $\sin\theta$ because

$$\tan\frac{\theta}{2} = \frac{\sin\frac{\theta}{2}}{\cos\frac{\theta}{2}} \text{ and } \sin\theta = 2\sin\frac{\theta}{2}\cos\frac{\theta}{2}.$$

Thus, we can remove the squares. The second identity for $\tan\frac{\theta}{2}$ is equivalent to the first because

$$\frac{\sin\theta}{1+\cos\theta} = \frac{1-\cos\theta}{\sin\theta} \iff \sin^2\theta = 1-\cos^2\theta,$$

which is the main Pythagorean identity. □

Problem 8.18. Prove the triple angle identities:

$$\sin(3\theta) = 3\sin\theta - 4\sin^3\theta,$$
$$\cos(3\theta) = 4\cos^3\theta - 3\cos\theta.$$

Can you use these to develop a formula for $\tan(3\theta)$ in terms of $\tan\theta$?

Theorem 8.19. The final set of identities allow us to convert between a sum or difference and a product.

Product-to-Sum	Sum-to-Product
$\sin(\alpha+\beta)+\sin(\alpha-\beta) = 2\sin\alpha\cos\beta$	$\sin\alpha+\sin\beta = 2\sin\left(\frac{\alpha+\beta}{2}\right)\cos\left(\frac{\alpha-\beta}{2}\right)$
$\sin(\alpha+\beta)-\sin(\alpha-\beta) = 2\cos\alpha\sin\beta$	$\sin\alpha-\sin\beta = 2\sin\left(\frac{\alpha-\beta}{2}\right)\cos\left(\frac{\alpha+\beta}{2}\right)$
$\cos(\alpha-\beta)+\cos(\alpha+\beta) = 2\cos\alpha\cos\beta$	$\cos\alpha+\cos\beta = 2\cos\left(\frac{\alpha+\beta}{2}\right)\cos\left(\frac{\alpha-\beta}{2}\right)$
$\cos(\alpha-\beta)-\cos(\alpha+\beta) = 2\sin\alpha\sin\beta$	$\cos\alpha-\cos\beta = -2\sin\left(\frac{\alpha+\beta}{2}\right)\sin\left(\frac{\alpha-\beta}{2}\right)$

Proof. The product-to-sum identities follow from expanding the two terms on the left side of each identity using the angle sum and angle difference identities. The sum-to-product identities follow from the product-to-sum identities. For illustrative purposes, we provide one of the proofs, and the rest is left to the reader: Suppose α and β are angles. Let $x = \frac{\alpha+\beta}{2}$ and $y = \frac{\alpha-\beta}{2}$. Then $x+y=\alpha$ and $x-y=\beta$. Therefore,

$$\begin{aligned}\sin\alpha+\sin\beta &= \sin(x+y)+\sin(x-y)\\ &= 2\sin x\cos y = 2\sin\left(\frac{\alpha+\beta}{2}\right)\cos\left(\frac{\alpha-\beta}{2}\right).\end{aligned}$$

□

Altogether, we have discussed over fifty trigonometric identities, if we include the variations in signs. It is not feasible to memorize all of them. We recommend memorizing the angle sum and difference identities, and using them to derive the others, as outlined. In addition, the values of trigonometric functions on axis angles like 0, $\frac{\pi}{2}$, π, and 2π (even $\frac{3\pi}{2}$ sometimes) can be used to obtain the reflection and phase shift identities as special cases. For example,

$$\sin\left(\theta+\frac{\pi}{2}\right) = \sin\theta\cos\frac{\pi}{2} + \sin\frac{\pi}{2}\cos\theta = \cos\theta$$

because $\cos\frac{\pi}{2} = 0$ and $\sin\frac{\pi}{2} = 1$. In the process of using identities, the more frequently occurring ones are automatically memorized in due time.

Chapter 9

Complex Numbers

"Between two truths of the real domain, the easiest and shortest path quite often passes through the complex domain."

– *Paul Painlevé*

"Schrödinger put the square root of minus one into the equation, and suddenly it made sense... And that square root of minus one means that nature works with complex numbers and not with real numbers."

– *Freeman Dyson, Birds and Frogs*

The complex numbers are a two-dimensional extension of real numbers, and surprisingly form a field. They have significant algebraic ramifications, such as in the theory of polynomials. We will start off by studying the rectangular form of complex numbers and follow it up with the polar form. We will then observe that complex numbers are essentially Cartesian coordinates with an extra multiplicative structure, which we will see has a geometric interpretation that involves rotation.

9.1 Rectangular Form

Definition 9.1. A **complex number** is an expression in the **rectangular form** $a+bi$, where a and b are real numbers, and i is a solution to the equation

$$x^2+1=0.$$

Note that, since the squares of all real numbers are non-negative, whereas $i^2=-1$, it must be true that i is distinct from all real numbers. The set of complex numbers is denoted by the symbol $\mathbb{C}$. The complex number $a+bi$ corresponds with the Cartesian coordinates (a,b). So, $i=0+1i$ corresponds with $(0,1)$.

Definition 9.2. For a complex number $a+bi$, the **real part** is $\mathrm{Re}(a+bi)=a$ and the **imaginary part** is $\mathrm{Im}(a+bi)=b$ (note that the imaginary part is the real number b, not bi). The real part and the imaginary parts are called the **components** of the complex number. Two complex numbers are equal if and only if their real parts are equal and their imaginary parts are equal.

Definition 9.3. For a complex number $z=a+bi$, if $b=0$ then z maybe be interpreted as the real number a. As such, $\mathbb{R}$ is a subset of $\mathbb{C}$. If instead $a=0$, then we call the complex number $z=bi$ **pure imaginary**.

Definition 9.4. We define **addition** and **multiplication** on complex numbers in way that, if the two complex numbers in question are real, the definitions of these operations on $\mathbb{C}$ boil down to the definitions of the operations on $\mathbb{R}$:

- Addition is component-wise, so that

$$(a+bi)+(c+di)=(a+c)+(b+d)i.$$

- Multiplication respects the distributive law, so that

$$\begin{aligned}(a+bi)\cdot(c+di)&=ac+adi+bci+bdi^2\\&=ac+adi+bci-bd\\&=(ac-bd)+(ad+bc)i.\end{aligned}$$

Theorem 9.5. The complex numbers, with addition and multiplication, form a field (fields are defined in Definition 3.14). To reiterate in an expanded fashion, this means:

1. Addition and multiplication are commutative.
2. Addition and multiplication are associative.
3. Multiplicative is distributive over addition.
4. There exists an additive identity and a multiplicative identity.
5. Every element has an additive inverse.
6. Every element that is not the additive identity has a multiplicative inverse.

By Theorem 3.8 and Theorem 3.10, identities and inverses in a field are unique.

Proof. We provide the proofs and ideas for some of the properties. The details are left to the reader.

4. It can be verified that $0 = 0+0i$ is an additive identity, and that $1 = 1+0i$ is a multiplicative identity.
5. It can be verified that the negative $-1 \cdot (a+bi) = (-a) + (-b)i$ is an additive inverse of $a+bi$. We denote the additive inverse of $a+bi$ by $-(a+bi)$.
6. Finding a multiplicative inverse of $a+bi \neq 0$ requires slightly more effort. Suppose $c+di$ exists such that $(c+di)(a+bi) = 1$. The critical observation is that $(a+bi)(a+(-b)i)$ expands to the real number a^2+b^2. So multiplying both sides of the equation by $a+(-b)i$ yields

$$(c+di)(a^2+b^2) = a+(-b)i \implies c+di = \frac{a}{a^2+b^2} + \frac{-b}{a^2+b^2}i.$$

It can then be manually verified that this is indeed a multiplicative inverse. We denote the multiplicative inverse of $a+bi$ by $(a+bi)^{-1}$.

□

Definition 9.6. Due to the existence of additive inverses for all complex numbers and multiplicative inverses for all non-zero complex numbers, we can define the inverse operations of **subtraction** and **division** on $\mathbb{C}$:

- Subtraction by a complex number $c+di$ is the addition of its additive inverse:

$$(a+bi)-(c+di) = (a-c)+(b-d)i.$$

- Division by a non-zero complex number $c+di$ is multiplication by its multiplicative inverse:

$$\frac{a+bi}{c+di} = (a+bi)(c+di)^{-1}.$$

Definition 9.7. We can define **powers** of a complex number z, as long as we only allow integer exponents n for the time being:

- If $n=0$ and $z \neq 0$, then $z^0 = 1$.
- If n is a positive integer, then $z^n = \underbrace{z \cdot z \cdots z}_{n \text{ copies of } z}$.
- If n is a negative integer then $z^n = (z^{-1})^{-n} = \underbrace{z^{-1} \cdot z^{-1} \cdots z^{-1}}_{(-n) \text{ copies of } z^{-1}}$.

Theorem 9.8. The usual exponents rules hold for complex numbers z and w, and integer exponents n and m, assuming we never take a base of 0 to the power of a negative exponent or divide by 0:

1. Negative exponent means reciprocal: $z^{-n} = \frac{1}{z^n} = (z^n)^{-1}$
2. Multiplying powers with the same base: $z^n \cdot z^m = z^{n+m}$
3. Dividing powers with the same base: $\frac{z^n}{z^m} = z^{n-m}$
4. Multiplying powers with the same exponent: $z^n \cdot w^n = (zw)^n$
5. Dividing powers with the same exponent: $\frac{z^n}{w^n} = \left(\frac{z}{w}\right)^n$
6. Nested powers: $(z^n)^m = z^{nm}$

Problem 9.9. In Example 3.51, we saw that the powers $(-1)^n$ oscillate between 1 and -1 for positive integers n. Find a pattern in the powers of i.

As mentioned earlier, each complex number $a+bi$ can be interpreted as a point (a, b) on the Cartesian plane, and vice versa. This allows us to see complex numbers visually, similar to how we can see real numbers on the number line. Addition of complex numbers is analogous to the addition of coordinates, as they are both component-wise. As we will see, the true strength of this new interpretation is the fact that complex numbers can be multiplied, which is a definition that we did not have for vanilla coordinates.

Definition 9.10. When finding the inverse of $z = a + bi$, we found the complex number number $a - bi$ to be useful. Denoted by

$$\overline{z} = a - bi,$$

we call it the **complex conjugate** of $a + bi$. Geometrically, the function $z \mapsto \overline{z}$ is a reflection across the x-axis. As a side note, the function that is the reflection across the y-axis is given by $z \mapsto -\overline{z}$.

Theorem 9.11. The complex conjugate has several useful and interesting properties. If z and w are complex numbers, a is a real number, and n is an integer then:

1. Conjugate of a real number: $\overline{z} = z$ if and only if $z = a$ for some real number a
2. Conjugate of a pure imaginary number: $\overline{z} = -z$ if and only if $z = ai$ for some real number a
3. Real and imaginary parts: $\text{Re}(z) = \frac{z + \overline{z}}{2}$ and $\text{Im}(z) = \frac{z - \overline{z}}{2i}$
4. Conjugation is an involution: $\overline{(\overline{z})} = z$
5. Conjugation distributes over addition: $\overline{z + w} = \overline{z} + \overline{w}$
6. Conjugation distributes over multiplication: $\overline{z \cdot w} = \overline{z} \cdot \overline{w}$
7. Conjugation commutes with taking the negative: $\overline{(-z)} = -\overline{z}$, and so $\overline{z - w} = \overline{z} - \overline{w}$
8. Conjugation commutes with taking the inverse: $(\overline{z})^{-1} = \overline{(z^{-1})}$, and so $\overline{\left(\frac{w}{z}\right)} = \frac{\overline{w}}{\overline{z}}$ for $z \neq 0$
9. Conjugation commutes with taking powers: $(\overline{z})^n = \overline{z^n}$

These are straightforward to verify, so their proofs are left to the reader.

Definition 9.12. During the process of finding the inverse of $z = a + bi$, the expression $a^2 + b^2$ also came up. We call the real number

$$|a + bi| = \sqrt{a^2 + b^2}$$

the **modulus** or **magnitude** or **absolute value** of $a + bi$. The reason that we take the square root of it is that the final expression then has a geometric interpretation. If we draw the line segment between $a + bi$ and 0, along with the perpendicular distance from $a + bi$ to the x-axis, then we have produced a right triangle with legs of length $|a|$ and $|b|$ (these are real absolute values). The hypotenuse is the distance between $a + bi$ and the origin, and the Pythagorean theorem tells us that it equals $\sqrt{a^2 + b^2} = |a + bi|$.

Theorem 9.13. The complex modulus has some convenient properties. If z and w are complex numbers, and n is an integer then:

1. When applied to complex numbers with 0 imaginary part, the complex modulus restricts to becoming the real absolute value.

2. Zero modulus: $|z| = 0$ if and only if $z = 0$

3. Modulus and conjugation: $z \cdot \overline{z} = |z|^2$; in particular, if z lies on the complex unit circle, meaning $|z| = 1$, then $\overline{z} = \frac{1}{z}$

4. Modulus is preserved under reflections: $|z| = |-z| = |\overline{z}| = \left|\overline{(-z)}\right|$

5. Modulus distributes over multiplication: $|z \cdot w| = |z| \cdot |w|$

6. Modulus commutes with taking the inverse: $|z^{-1}| = |z|^{-1}$, and so $\left|\frac{w}{z}\right| = \frac{|w|}{|z|}$ for $z \neq 0$

7. Modulus commutes with taking powers: $|z^n| = |z|^n$

8. Real and imaginary parts bounds: $|z| \geq |\mathrm{Re}(z)|$ and $|z| \geq |\mathrm{Im}(z)|$

9. **Complex triangle inequality**: $|z| + |w| \geq |z + w|$; equality holds if and only if $z = 0$ or $w = 0$ or $z = \alpha w$ for some real $\alpha > 0$, meaning z and w fall on the same ray emanating from the origin

Proof. We leave all but the triangle inequality to the reader to verify, as the proofs are not difficult. In order to prove this inequality, we will use the substitutions $z = a + bi$ and $w = c + di$, and work backwards through repeated squaring and simplification:

$$
\begin{aligned}
|z| + |w| \geq |z + w| &\iff |a + bi| + |c + di| \geq |a + bi + c + di| \\
&\iff \sqrt{a^2 + b^2} + \sqrt{c^2 + c^2} \geq \sqrt{(a + c)^2 + (b + d)^2} \\
&\iff a^2 + b^2 + c^2 + d^2 + 2\sqrt{(a^2 + b^2)(c^2 + d^2)} \geq a^2 + c^2 + b^2 + d^2 + 2ac + 2bd \\
&\iff \sqrt{(a^2 + b^2)(c^2 + d^2)} \geq ac + bd.
\end{aligned}
$$

This actually follows from the two-dimensional case of the Cauchy-Schwarz inequality (see the proof of Theorem 12.9), but, for now, we will prove it by expansion.

$$
\begin{aligned}
\sqrt{(a^2 + b^2)(c^2 + d^2)} \geq ac + bd &\impliedby \sqrt{(a^2 + b^2)(c^2 + d^2)} \geq |ac + bd| \geq ac + bd \\
&\iff (a^2 + b^2)(c^2 + d^2) \geq (ac + bd)^2 \\
&\iff a^2c^2 + a^2d^2 + b^2c^2 + b^2d^2 \geq a^2c^2 + b^2d^2 + 2abcd \\
&\iff a^2d^2 + b^2c^2 \geq 2abcd \\
&\iff (ad - bc)^2 \geq 0,
\end{aligned}
$$

which follows from the trivial inequality.
Equality holds if and only if $ad = bc$ and $|ac + bd| = ac + bd$. It is easy to verify that, if $z = 0$ or $w = 0$ or $z = \alpha w$ for some positive real α, then the above equality condition is implied. So we must prove the other direction. Suppose $ad = bc$ and $|ac + bd| = ac + bd$, and that $z \neq 0$ and $w \neq 0$. By some basic linear algebra from Volume 4, $ad = bc$ means that the position vectors with arrowheads (a, b) and (c, d) are linearly dependent, which means they must lie on the same line through the origin. By an equivalent criterion for linear dependence from Volume 4, since $z \neq 0$, there exists a real α such that $z = \alpha w$. All we need to do now is prove that they lie on the same ray through the origin, or, equivalently, they point in the same direction. To do this, it suffices to prove that $\alpha > 0$. Indeed, by $|ac + bd| = ac + bd$, we get

$$
\begin{aligned}
|ac + bd| &= |\alpha c \cdot d + \alpha d \cdot d| = |\alpha| \cdot (c^2 + d^2), \\
ac + bd &= \alpha c \cdot d + \alpha d \cdot d = \alpha \cdot (c^2 + d^2),
\end{aligned}
$$

so $|\alpha| = \alpha$, making α non-negative. Finally, $\alpha \neq 0$ because $\alpha = 0$ would imply $z = 0$, which is a contradiction. Therefore, α is strictly positive. □

Theorem 9.14 (Complex geometric series). Given any non-negative integer t and complex number $z \neq 1$,

$$1 + z + z^2 + \cdots + z^t = \frac{z^{t+1} - 1}{z - 1}.$$

Proof. The proof is the same as that for summing a geometric series with real summands (Theorem 7.7). Let

$$S = 1 + z + z^2 + \cdots + z^t.$$

By multiplying both sides by z, we get

$$zS = z + z^2 + z^3 + \cdots + z^{t+1}.$$

Subtracting the latter equation from the former yields $(1 - z)S = 1 - z^{t+1}$. Therefore, $S = \dfrac{1 - z^{t+1}}{1 - z}$. □

Problem 9.15. Prove the following variant of the formula for a finite geometric series. Let $z \neq 1$ be a complex number. Let α and β be positive integers and γ be a non-negative integer such that $\gamma = \alpha - q\beta$ for some non-negative integer q. Then

$$z^{\alpha} + z^{\alpha-\beta} + z^{\alpha-2\beta} + \cdots + z^{\gamma} = \frac{z^{\alpha+\beta} - z^{\gamma}}{z^{\beta} - 1}.$$

9.2 Polar Form

So far, we have been working with complex numbers in the rectangular form $a + bi$. There is another way of representing complex numbers, which illuminates further properties.

Definition 9.16. Instead of locating points $z \neq 0$ on the Cartesian plane using rectangular coordinates (x, y), we can use **polar coordinates**. Polar coordinates also have two components (r, θ). The first component $r \in (0, \infty)$ represents the circle centred at the origin on which z lies; we compute r as the length of the radius from z to the origin, which is the modulus $|z|$. The second component $\theta \in [0, 2\pi)$ is the angle by which the positive x-axis has to be rotated counterclockwise in order to coincide with the aforementioned radius. These coordinates $(r, \theta) \in (0, \infty) \times [0, 2\pi)$ are unique representations of points on the plane, and so are in bijection with rectangular coordinates $(x, y) \in \mathbb{R} \times \mathbb{R} \backslash \{(0, 0)\}$. If we interpret z as a complex number, then we denote and define the **argument** of z as $\arg z = \theta$. Similar to rectangular coordinates, two complex numbers are equal if and only if their moduli are equal and their arguments are equal.

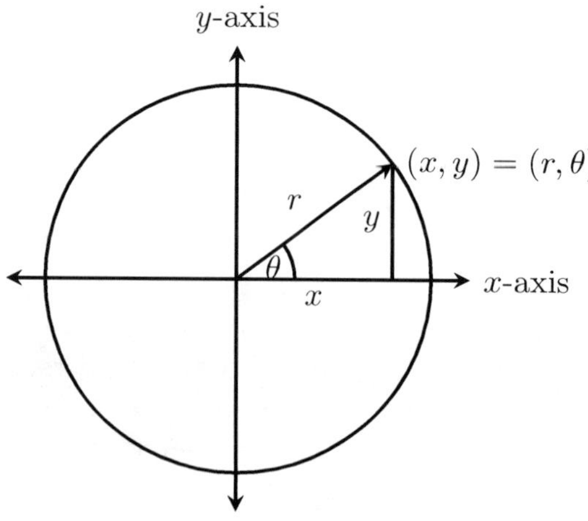

Now, we will tackle the question of how to convert between rectangular and polar coordinates.

Theorem 9.17. Suppose $z \neq 0$ is a point on the Cartesian plane. If the polar coordinates for z are (r, θ), then the corresponding rectangular coordinates (x, y) are

$$x = r \cdot \cos\theta,$$
$$y = r \cdot \sin\theta.$$

Proof. Let $x' = \cos\theta$ and $y' = \sin\theta$. Then (x', y') is the point on the unit circle that makes an angle of θ with the positive x-axis, by the definition of sine and cosine. The radius from the origin to (x', y') has length 1. By dilating the unit circle from the origin by a factor of r, (x', y') is mapped to (rx', ry'), which coincides with z. Thus,

$$z = (rx', ry') = (r \cdot \cos\theta, r \cdot \sin\theta).$$

This use of dilation can be made formal using similar triangles, which we leave to the reader. □

Theorem 9.18. Suppose $z \neq 0$ is a point on the Cartesian plane. If the rectangular coordinates for x are (x, y), then the corresponding polar coordinates (r, θ) are

$$r = \sqrt{x^2 + y^2},$$
$$\theta = \begin{cases} \arccos \frac{x}{r} & \text{if } y \geq 0 \\ 2\pi - \arccos \frac{x}{r} & \text{if } y < 0 \end{cases}.$$

Authors often present a formula for θ using arctan but we have preferred to express θ using arccos because it involves fewer cases.

Proof. The fact that $r = \sqrt{x^2 + y^2}$ follows from the fact that r is the distance from (x, y) to 0. To compute θ, we will tackle the case $y \geq 0$m followed by the case $y < 0$. In both cases, we will use the fact that Definition 8.12 says that the range of arccos is $[0, \pi]$.

- Suppose $y \geq 0$. By observing the unit circle, this means $\theta \in [0, \pi]$. We know that $x = r \cdot \cos\theta$. Since $\theta \in \mathrm{Rng}(\arccos)$, we can isolate θ to get

$$\theta = \arccos \frac{x}{r}.$$

- Now suppose $y < 0$. By observing the unit circle, this means $\pi < \theta < 2\pi$ which is equivalent to $0 < 2\pi - \theta < \pi$. Using the period of cosine and a reflection identity,

$$\cos(2\pi - \theta) = \cos(-\theta) = \cos\theta.$$

As before, $\cos\theta = \frac{x}{r}$, so $\cos(2\pi - \theta) = \frac{x}{r}$. Since $(2\pi - \theta) \in \mathrm{Rng}(\arccos)$, we can isolate θ to get

$$\theta = 2\pi - \arccos \frac{x}{r}.$$

□

Definition 9.19. We defined $\arg(z)$ to lie in $[0, 2\pi)$, but it often suffices to use any value of the set

$$\mathrm{Arg}(z) = \{\arg(z) + 2\pi k : k \in \mathbb{Z}\}$$

because, if combined with a fixed modulus, they all lead to the same point in the plane. We say that angles α and β are **congruent modulo** 2π if $\alpha - \beta = 2\pi k$ for some integer k, and we denote this relationship by $\alpha \equiv \beta$.

Lemma 9.20. If $\alpha = \beta$ then of course $\alpha \equiv \beta$. The converse is not necessarily true, but we can establish equality using congruence if the angles lie within the same revolution. To be precise, if we know that $\alpha \equiv \beta$ then $\alpha = \beta$ if either of the following criteria hold:

1. Both α and β lie in the same interval $[x, y)$ or $(x, y]$ or (x, y), where $0 < y - x \leq 2\pi$.

2. Both α and β lie in the same interval $[x, y]$, where $0 \leq y - x < 2\pi$.

The closed and open nature of an endpoint of the interval is important, as is the strictness or non-strictness of the bounds of the inequalities.

Proof. We provide a proof for the interval $[x, y)$ in the first case and leave the rest as an exercise in mimicry to the reader. As both α and β lie in $[x, y)$, we have the inequalities

$$x \leq \alpha < y,$$
$$x \leq \beta < y \implies -y < -\beta \leq -x.$$

Their sum is $x - y < \alpha - \beta < y - x$ or

$$0 \leq |\alpha - \beta| < y - x \leq 2\pi.$$

Since $\alpha \equiv \beta$, we know that $\alpha - \beta$ is an integer multiple of 2π. The only multiple of 2π in $[0, 2\pi)$ is 0, which completes the proof. □

Definition 9.21. Let $z = x + iy$ be a non-zero element of the complex plane. By the existence of polar coordinates, we know that there exists an ordered pair $(r, \theta) \in (0, \infty) \times [0, 2\pi)$ such that $x = r \cdot \cos\theta$ and $y = r \cdot \sin\theta$. This means

$$z = x + iy = r(\cos\theta + i\sin\theta).$$

This is called the **trigonometric form** of a complex number. Note that we may replace θ with any angle congruent to θ modulo 2π without changing the underlying point in the plane. So, even though we call it "the" trigonometric form, this form is not unique. As a shorthand that has deeper meaning, we define

$$re^{i\theta} = r(\cos\theta + i\sin\theta),$$

where e is a real number called Euler's constant.

It can be shown that is definition is consistent with the infinite series expressions for $e^{i\theta}$, $\cos\theta$, and $\sin\theta$ but we will not delve into this. Instead, we will demonstrate that this exponential notation is appropriate by proving several analogues of exponent laws for expressions of the form $e^{i\theta}$.

Theorem 9.22. The follow exponent laws hold for any real α and β, positive real r and s, and any integer n:

1. $re^{i\alpha} \cdot se^{i\beta} = rse^{i(\alpha+\beta)}$

2. $(re^{i\alpha})^{-1} = \frac{1}{r}e^{i(-\alpha)}$ and so $\overline{re^{i\theta}} = re^{i(-\theta)}$

3. $\frac{re^{i\alpha}}{se^{i\beta}} = \frac{r}{s}e^{i(\alpha-\beta)}$

4. $(re^{i\alpha})^n = r^n e^{i(n\alpha)}$ (de Moivre's formula)

Proof. We prove these identities one by one.

1. By the sine and cosine sum identities (Theorem 8.16),

$$\begin{aligned}
re^{i\alpha} \cdot se^{i\beta} &= r(\cos\alpha + i\sin\alpha) \cdot s(\cos\beta + i\sin\beta) \\
&= rs\left((\cos\alpha\cos\beta - \sin\alpha\sin\beta)\right) + irs\left((\sin\alpha\cos\beta + \cos\alpha\sin\beta)\right) \\
&= rs\left(\cos(\alpha+\beta) + i\sin(\alpha+\beta)\right) \\
&= rse^{i(\alpha+\beta)}.
\end{aligned}$$

2. Using the fact that $(a+bi)^{-1} = \dfrac{a}{a^2+b^2} + i\dfrac{-b}{a^2+b^2}$ and the Pythagorean identity and reflection identities,

$$\begin{aligned}(re^{i\alpha})^{-1} &= (r\cos\alpha + ir\sin\alpha)^{-1}\\ &= \frac{r\cos\alpha}{r^2(\cos^2\alpha+\sin^2\alpha)} + i\frac{-r\sin\alpha}{r^2(\cos^2\alpha+\sin^2\alpha)}\\ &= \frac{1}{r}(\cos\alpha - i\sin\alpha)\\ &= \frac{1}{r}(\cos(-\alpha)+i\sin(-\alpha))\\ &= \frac{1}{r}e^{i(-\alpha)}.\end{aligned}$$

As a consequence, since $(re^{i\theta})\cdot\overline{(re^{i\theta})} = r^2$ (due to $z\cdot\overline{z} = |z|^2$, mentioned in Theorem 9.13),

$$\overline{(re^{i\theta})} = r^2\cdot(re^{i\theta})^{-1} = re^{i(-\theta)}.$$

3. Using the previous two parts,

$$\frac{re^{i\alpha}}{se^{i\beta}} = (re^{i\alpha})\cdot(se^{i\beta})^{-1} = re^{i\alpha}\cdot\frac{1}{s}e^{i(-\beta)} = \frac{r}{s}e^{i(\alpha-\beta)}.$$

4. To prove de Moivre's formula, we will split it into the cases $n=0, n\geq 1$, and $n\leq -1$. For $n=0$,

$$(re^{i\theta})^0 = 1 = r^0(\cos 0 + i\sin 0) = r^0e^{i0} = r^0e^{i(0\theta)}.$$

For $n\geq 1$, this is an induction exercise using

$$re^{i\alpha}\cdot se^{i\beta} = rse^{i(\alpha+\beta)}.$$

If $n\leq -1$, let $m=-n\geq 1$. By previous parts,

$$(re^{i\theta})^n = (re^{i\theta})^{(-1)\cdot m} = \left((re^{i\theta})^m\right)^{-1} = (r^me^{i(m\theta)})^{-1} = r^{-m}e^{i(-m\theta)} = r^ne^{i(n\theta)}.$$

□

Problem 9.23. Prove that

$$\sin\theta = \frac{e^{i\theta}-e^{i(-\theta)}}{2i} \text{ and } \cos\theta = \frac{e^{i\theta}+e^{i(-\theta)}}{2}.$$

Lemma 9.24. We now establish some results about trigonometric forms.

1. For all real θ, $e^{i\theta} = \cos\theta + i\sin\theta = 1$ if and only if $\theta\equiv 0$.
2. For all real α and β, $e^{i\alpha} = e^{i\beta}$ if and only if $\alpha\equiv\beta$.
3. For all real θ, $\arg(e^{i\theta})\equiv\theta$.
4. For all real α and β, $\arg(e^{i\alpha}) = \arg(e^{i\beta})$ if and only if $\alpha\equiv\beta$.

Proof. We will prove each part in succession, using earlier parts to prove later parts.

1. Suppose $e^{i\theta} = \cos\theta + i\sin\theta = 1$. Then we can equate real and imaginary parts to get $\cos\theta = 1$ and $\sin\theta = 0$, which simultaneously hold only at angles that are integer multiples of 2π. So $\theta\equiv 0$.
2. Subsequently, we have the following equivalences:

$$e^{i\alpha} = e^{i\beta} \iff \frac{e^{i\alpha}}{e^{i\beta}} = 1 \iff e^{i(\alpha-\beta)} = 1 \iff \alpha-\beta\equiv 0 \iff \alpha\equiv\beta.$$

3. Let $e^{i\theta} = x + iy$ and $\arg(x + iy) = \phi$. We know that $x = \cos\phi$ and $y = \sin\phi$. Then
$$e^{i\theta} = x + iy = \cos\phi + i\sin\phi = e^{i\phi}.$$
By the previous part, this means $\arg(e^{i\theta}) = \phi \equiv \theta$.

4. This follows from the previous two parts. If
$$\arg(e^{i\alpha}) = \arg(e^{i\beta}),$$
then, using the fact, that
$$\arg(e^{i\alpha}) \equiv \alpha$$
$$\arg(e^{i\beta}) \equiv \beta,$$
we get $\alpha \equiv \beta$. Conversely, if $\alpha \equiv \beta$, then we know that $e^{i\alpha} = e^{i\beta}$. Taking the argument of each sides yields
$$\arg(e^{i\alpha}) = \arg(e^{i\beta}).$$

□

Corollary 9.25. For any non-zero complex number z, we can redefine $\arg z$ to be the unique real number $\theta \in [0, 2\pi)$ such that $z = |z| \cdot e^{i\theta}$. This is a less geometric description of the argument than in Definition 9.16.

Proof. Suppose z is a non-zero complex number with modulus $r = |z|$ and $\arg z = \theta$. We know that $z = re^{i\theta}$ is a trigonometric form of z. Suppose z has another trigonometric form $se^{i\phi}$ for some $s > 0$ and $\phi \in [0, 2\pi)$. Then $r = |z| = |se^{i\phi}| = s$ and $e^{i\phi} = e^{i\theta}$. By Lemma 9.24, this means $\phi \equiv \theta$. But they both lie in $[0, 2\pi)$ which forces $\phi = \theta$. Therefore, $\arg z$ is the unique real number $\theta \in [0, 2\pi)$ such that $z = |z| \cdot e^{i\theta}$. □

Corollary 9.26. Let z and w be non-zero complex numbers and n be an integer. Then the following analogues of logarithmic rules (Theorem 3.60) hold for arg modulo 2π:

1. $\arg(zw) \equiv \arg z + \arg w$
2. $\arg(z^{-1}) \equiv -\arg z$
3. $\arg\left(\frac{w}{z}\right) \equiv \arg w - \arg z$
4. $\arg(z^n) \equiv n\arg(z)$
5. $\arg z = 0$ if and only if z is a positive real number, and $\arg z = \pi$ if and only if z is a negative real number

Proof. We show how to prove the first one and leave the rest to the reader to verify. We will use the fact that $\arg(rz) = \arg(z)$ for any positive real r, which is straightforward to see. We will also use the various analogues of exponent laws that we have proven for complex numbers in trigonometric form (Lemma 9.24), along with the fact that $\arg(e^{i\theta}) \equiv \theta$. Let $z = re^{i\alpha}$ and $w = se^{i\beta}$ where $\alpha = \arg z$ and $\beta = \arg w$. Then
$$\arg(zw) = \arg(re^{i\alpha} \cdot se^{i\beta}) = \arg(e^{i\alpha} \cdot e^{i\beta}) = \arg(e^{i(\alpha+\beta)}) \equiv \alpha + \beta = \arg z + \arg w.$$

□

The following corollary establishes an important link between the polar form and geometry, specifically by interpreting multiplication by a complex number as a combination of dilation and rotation.

Corollary 9.27. The function $s_w : \mathbb{C} \to \mathbb{C}$, defined by $s_w(z) = zw$ has a geometric interpretation on the complex plane. It means we apply a positive dilation to z from the origin by a factor of $|w|$ and rotate it counterclockwise around the origin by an angle measuring $\arg w$. This is called a **spiral similarity**. In the specific case where w lies on the complex unit circle, meaning $|w| = 1$, then this function causes purely a rotation without a dilation.

Proof. The geometric interpretation follows from the following two facts that we have previously stated:

$$|zw| = |z| \cdot |w|,$$
$$\arg(zw) \equiv \arg z + \arg w.$$

The special case true by the definition, $|w| = 1$. □

Problem 9.28. Determine a complex number w such that multiplying any complex number z by w results in z being rotated counter-clockwise by $\frac{\pi}{2} = 90°$.

Problem 9.29. Let z be a non-zero complex number. Then prove the following rules for arg:

$$\arg(-z) \equiv \pi + \arg z,$$
$$\arg(\overline{z}) \equiv -\arg z.$$

As a precursor to the fundamental theorem of algebra (Theorem 11.28), which tells us how many complex roots a polynomial with complex coefficients has, we will now compute all the nth roots of a non-zero complex number in trigonometric form.

Theorem 9.30. Any non-zero complex number z has exactly n distinct n^{th} roots. Since we have not actually defined n^{th} roots of complex numbers, this is made more precise in the language of equations by saying that there are exactly n distinct solutions $x = w$ of the equation $x^n - z$, where x is a variable and z is a fixed non-zero complex number. Moreover, if z is given in trigonometric form, we can compute each root w in trigonometric form.

Proof. Let $z = re^{i\theta}$. Then $w = se^{i\phi}$ is an nth root of z if and only if

$$re^{i\theta} = (se^{i\phi})^n = s^n e^{i(n\phi)}.$$

This holds if and only if $s = \sqrt[n]{r}$ and $\theta \equiv n\phi$. The former condition is an explicit statement about the modulus of w. We will now focus on the latter condition. The condition $\theta \equiv n\phi$ is equivalent to there existing an integer k such that $\theta + 2\pi k = n\phi$, which we can write as

$$\phi = \frac{\theta + 2\pi k}{n}.$$

This would seem to imply that there is a value of ϕ for each integer k, but it possible that many such angles lead to the same point in the plane. We need to classify such overlaps. This possibility is encapsulated by teh equivalent statements

$$se^{\frac{\theta+2\pi k_1}{n}} = se^{\frac{\theta+2\pi k_2}{n}} \iff \frac{\theta + 2\pi k_2}{n} \equiv \frac{\theta + 2\pi k_2}{n},$$

which are further equivalent to the existence of an integer j such that

$$\frac{\theta + 2\pi k_2}{n} + 2\pi j = \frac{\theta + 2\pi k_2}{n}.$$

Upon simplification, this is equivalent to $k_1 \equiv k_2 \pmod{n}$. As there are n distinct classes modulo n (to learn about modular arithmetic, see Volume 3), corresponding to $k = 0, 1, 2, \ldots, n-1$, each such k leads to a distinct root

$$w_k = \sqrt[n]{r}e^{\frac{\theta+2\pi k}{n}}.$$

As a geometric interpretation, notice that

$$\frac{w_{k+1}}{w_k} = e^{i\frac{2\pi}{n}},$$

where we define $w_n = w_0$. This means that each root can be found by rotating the preceding root around the origin by $\frac{2\pi}{n}$ radians. Subsequently, the n roots form a regular n-gon centred at the origin. □

Students who want to learn more about the usage of complex numbers in geometry, especially in olympiad problems, should refer to Evan Chen's acclaimed EGMO book [5].

Definition 9.31. The nth **roots of unity** are the n complex n^{th} roots of 1. They are given by

$$e^{i\frac{2\pi k}{n}}, \text{ for } k = 0, 1, 2, \ldots, n-1.$$

We will study these in Volume 3 in the context of number theory, specifically cyclotomic polynomials.

Chapter 10

Quadratics

"[Taniyama] was not a very careful person as a mathematician. He made a lot of mistakes. But he made mistakes in a good direction. I tried to imitate him. But I've realized that it's very difficult to make good mistakes."

– *Goro Shimura*

We will move towards studying polynomials by first looking at quadratic functions. We will start by seeing techniques for determining the roots of quadratics, including completing the square, which produces the quadratic formula, and factoring when it is feasible. We will also look at the shape of the graph of a quadratic function, which can be studied by completing the square and applying the trivial inequality.

10.1 Algebra

After linear functions in one variable, the functions in one variable that are one degree higher in complexity are quadratic functions.

Definition 10.1. A **quadratic function** in one variable is a function $f : \mathbb{C} \to \mathbb{C}$ of the form

$$f(x) = ax^2 + bx + c,$$

where $a \neq 0$ (this prevents it from being linear). The **coefficients** a, b, c are (for now) real constants that are respectively called the **leading** or **quadratic coefficient**, **linear coefficient**, and **constant**. Similarly, the **terms** ax^2, bx, c are respectively called the **leading** or **quadratic term**, **linear term**, and **constant term**. By Definition 2.41, a **root** of a quadratic function f is a complex number z such that $f(z) = 0$. A **quadratic equation** is formed by setting a quadratic function equal to 0, which produces

$$ax^2 + bx + c = 0.$$

Solving for the roots of a quadratic function amounts to setting it equal to 0 as shown, and determining all complex values of the variable x that satisfy this equation.

Lemma 10.2. If z is a complex number, then we say that w is a complex **square root** of z if $z = w^2$. Suppose a is a real number.

1. If $a = 0$, then the only complex square root of a is 0.
2. If $a > 0$, then the only two complex square roots of a are the real numbers $\pm\sqrt{a}$.
3. If $a < 0$, then the only two complex square roots of a are the pure imaginary numbers $\pm i\sqrt{-a}$ (note that $-a$ is positive).

Proof. Suppose $p + iq$ is a complex square root of the real number a. Then,

$$a = (p + iq)^2 = (p + iq)(p + iq) = (p^2 - q^2) + i(2pq).$$

Equating real and imaginary parts yields $a = p^2 - q^2$ and $0 = 2pq$, where the latter implies that at least one of p or q is 0. Note that $(-p - iq)^2 = a$ as well, so the negative of a square root of a is also a square root of a. Since one is the negative of the other and we are interested in finding all square roots, we may assume without loss of generality in our hunt that $p \geq 0$. Now we will treat a depending on its sign:

1. Suppose $a = 0$. Due to $pq = 0$, at least one of p or q is 0. Since $a = 0$, we get $|p| = |q|$ from $0 = p^2 - q^2$, meaning both are 0. Thus, 0 is the only square root of 0.

2. Suppose $a > 0$. Both p and q cannot be 0 because then $a = p^2 - q^2$ would be 0. Moreover, p cannot be 0, otherwise it would be true that $a = -q^2 < 0$. Since at least one of p or q is 0, we have $q = 0$. Thus, $p = \sqrt{a}$ and there are exactly two square roots $\pm\sqrt{a}$ of $a > 0$.

3. Suppose $a < 0$. Again, p and q cannot both be 0. Moreover, q cannot be 0, otherwise it would be true that $a = p^2 > 0$. Since at least one of p or q is 0, we have $p = 0$. Thus, $q = \sqrt{-a}$ and there are exactly two square roots $\pm i\sqrt{-a}$ of $a < 0$.

□

Definition 10.3. For convenience of notation, if $a < 0$ then we may choose to denote $\pm i\sqrt{-a}$ by $\pm\sqrt{a}$.

Example. We recommend being very careful with expressions of the form $\sqrt{a}$ for negative a as normal exponent laws do not necessarily hold for them. One famous example of a mistake is

$$-1 = i \cdot i = \sqrt{-1}\sqrt{-1} = \sqrt{(-1)\cdot(-1)} = \sqrt{1} = 1.$$

Problem 10.4. Suppose $f(x) = ax^2 + bx + c$ is a quadratic function.

1. If $c = 0$, then find all complex roots of f in terms of a and b.

2. If $b = 0$, then find all complex roots of f in terms of a and c.

Now we will show a general method of determining the roots of a quadratic, called the quadratic formula, which may be found using a process called "completing the square."

Theorem 10.5 (Quadratic formula). Let $f(x) = ax^2 + bx + c$ be a quadratic function. Then f has exactly two roots, which are

$$z = \frac{-b \pm \sqrt{b^2 - 4ac}}{2a}.$$

Proof. The standard derivation of the quadratic formula involves a step that requires casework on $\pm$ signs. We present a slightly modified method that circumvents the need to address these technicalities. A complex number z is a root of f if and only if

$$\begin{aligned}
az^2 + bz + c &= 0 \\
4a^2z^2 + 4abz + 4ac &= 0 \\
4a^2z^2 + 4abz + b^2 &= b^2 - 4ac \\
(2az + b)^2 &= b^2 - 4ac \\
2az + b &= \pm\sqrt{b^2 - 4ac} \\
z &= \frac{-b \pm \sqrt{b^2 - 4ac}}{2a}.
\end{aligned}$$

Each step was reversible, so there are exactly two (possibly equal) roots. □

Note that, even though the quadratic formula asserts the existence of exactly two roots, one or both of the roots might be extraneous when there is additional context. This is because there can be further constraints in the conditions stated in a problem, such as negative values not being allowed in the x-axis, which might be interpreted as time.

Definition 10.6. The **discriminant** of a quadratic function $f(x) = ax^2 + bx + c$ is the real number $b^2 - 4ac$, which is the expression that appears under the square root in the quadratic formula.

Corollary 10.7. We can classify solution types of a quadratic function $f(x) = ax^2 + bx + c$ according to the sign of its discriminant $D = b^2 - 4ac$:

1. $D > 0$ if and only if the two roots are real and distinct.
2. $D = 0$ if and only if the two roots are real and equal. In this case,
$$f(x) = a\left(x + \frac{b}{2a}\right)^2.$$
3. $D < 0$ if and only if the two roots are non-real and distinct.

Proof. Since the discriminant of $f(x) = ax^2 + bx + c$ is the expression under the square root in the quadratic formula
$$x = \frac{-b \pm \sqrt{b^2 - 4ac}}{2a},$$
the classification of the solution types based on the sign of D follows immediately. In the $D = 0$ case, we can write the condition $b^2 - 4ac = 0$ as $c = \frac{b^2}{4a}$, which leads to
$$ax^2 + bx + c = ax^2 + bx + \frac{b^2}{4a} = a\left(x + \frac{b}{2a}\right)^2.$$
□

Problem 10.8. If the coefficients of the quadratic function
$$f(x) = ax^2 + bx + c$$
are all rational, prove that the two roots are rational if and only if the discriminant D is the square of a rational number.

Corollary 10.9. Let $f(x) = ax^2 + bx + c$ be a quadratic function. If its roots are z_1 and z_2, then
$$f(x) = a(x - z_1)(x - z_2).$$

Proof. By the quadratic formula and the difference of squares factorization,
$$\begin{aligned}
a(x - z_1)(x - z_2) &= a\left(x - \frac{-b + \sqrt{b^2 - 4ac}}{2a}\right)\left(x - \frac{-b - \sqrt{b^2 - 4ac}}{2a}\right) \\
&= a\left(\left(x + \frac{b}{2a}\right)^2 - \left(\frac{\sqrt{b^2 - 4ac}}{2a}\right)^2\right) \\
&= a\left(x^2 + \frac{b}{a}x + \frac{b^2}{4a^2} - \frac{b^2 - 4ac}{4a^2}\right) \\
&= a\left(x^2 + \frac{b}{a}x + \frac{c}{a}\right) \\
&= ax^2 + bx + c.
\end{aligned}$$
□

Corollary 10.10 (Vieta's formulas for quadratics)**.** Two complex numbers z_1 and z_2 are the roots of a quadratic function
$$f(x) = ax^2 + bx + c$$
if and only if the following two equations hold:
$$z_1 + z_2 = -\frac{b}{a},$$
$$z_1 z_2 = \frac{c}{a}.$$

Proof. Suppose z_1 and z_2 are the roots of f. By Corollary 10.9, it holds for all real x that

$$\begin{aligned} ax^2+bx+c &= f(x) \\ &= a(x-z_1)(x-z_2) \\ &= a(x^2-(z_1+z_2)x+z_1z_2) \\ &= ax^2-a(z_1+z_2)x+az_1z_2. \end{aligned}$$

Putting all of the terms on one side, we get

$$(b+a(z_1+z_2))x+(c-az_1z_2)=0$$

for all real x. Substituting in $x=0$ yields $z_1z_2=\frac{c}{a}$. Then substituting in $x=1$ yields $z_1+z_2=-\frac{b}{a}$. Conversely, suppose z_1 and z_2 are two complex numbers that satisfy $z_1+z_2=-\frac{b}{a}$ and $z_1z_2=\frac{c}{a}$. For all real numbers x,

$$a(x-z_1)(x-z_2)=ax^2-a(z_1+z_2)x+az_1z_2=ax^2+bx+c=f(x).$$

Substituting z_1 and z_2 into both sides yields $0=f(z_1)$ and $0=f(z_2)$, so z_1 and z_2 are the roots of f. □

Readers are warned that it is a common mistake to forget to divide by a in Vieta's formulas. This might occur when one becomes accustomed to the monic case of $a=1$. It is also easy to forget the negative sign in the formula for the sum of the roots.

Theorem 10.11. If the coefficients a,b,c of a quadratic function $f(x)=ax^2+bx+c$ are integers, then there are some practical steps that can be taken to find the roots of f without using the quadratic formula.

1. If $a=1$, then we can iterate through the pairs of integers (r,s) such that $rs=c$. If we happen to strike upon a pair (r,s) such that $r+s=-b$, then r and s are the roots of f. Subsequently,

$$f(x)=(x-r)(x-s).$$

2. More generally, even if $a\neq 1$, we can iterate through the pairs of integers (p,q) such that $pq=a$ and pairs of integers (r,s) such that $rs=c$. If we happen to strike upon pairs (p,q) and (r,s) such that $ps+qr=-b$, then $\frac{r}{p}$ and $\frac{s}{q}$ are the roots of f. Subsequently,

$$f(x)=(px-r)(qx-s).$$

Writing the quadratic function as a product of two linear functions is called **factorization** and the two linear functions are called **factors**. The technique can be extended to the case where some or all of the coefficients a,b,c are non-integer rationals, as we can set ax^2+bx+c equal to 0 and clear the denominators. Note that it is possible for the roots to not be rational even if the coefficients are rational, so it is not always possible to factor a quadratic into two linear factors that each have integer coefficients.

Proof. Let $f(x)=ax^2+bx+c$ be a quadratic function with integer coefficients a,b,c.

1. Suppose $a=1$. If r,s satisfy $rs=c$ and $r+s=-b$, then Vieta's formulas tell us that r and s are the roots of f. This leads to

$$f(x)=(x-r)(x-s).$$

2. If p,q and r,s satisfy

$$\begin{aligned} pq &= a, \\ ps+qr &= -b, \\ rs &= c, \end{aligned}$$

then dividing the second and third equations by a and using the first equation yields

$$\frac{s}{q} + \frac{r}{p} = -\frac{b}{pq} = -\frac{b}{a},$$
$$\frac{r}{p} \cdot \frac{s}{q} = -\frac{c}{pq} = -\frac{c}{a}.$$

By Vieta's formulas, $\frac{r}{p}$ and $\frac{s}{q}$ are the roots of f. This leads to

$$f(x) = a\left(x - \frac{r}{p}\right)\left(x - \frac{s}{q}\right) = (px - r)(qx - s).$$

□

A biconditional criterion for when quadratics can be factored, along with a practical method of implementing factoring quadratics, called "decomposition," is given in Volume 3.

Theorem 10.12. Let $z = x + iy$ be a non-real complex number. Then z has exactly two complex square roots, which are

$$\sqrt{\frac{\sqrt{x^2 + y^2} + x}{2}} + i \cdot \operatorname{sgn}(y)\sqrt{\frac{\sqrt{x^2 + y^2} - x}{2}}$$

and its negative.

Proof. Suppose a and b are real numbers such that $x + iy = (a + ib)^2$. Then it would also be true that $x + iy = (-a - ib)^2$. So, finding one root automatically leads to a second, where the two are negatives of each other. This allows us to assume without loss of generality in our search that, if such a complex number $a+ib$ exists, then $a \geq 0$. Note that a cannot be 0 because then it would be true that $x+iy = (a+ib)^2 = -b^2$, which is is real, contradicting the assumption that $z = x + iy$ is non-real. So $a > 0$.
Now we expand

$$x + iy = (a + ib)^2 = (a^2 - b^2) + 2abi$$

and equate real and imaginary parts to get

$$x = a^2 - b^2 \quad \text{and} \quad y = 2ab.$$

Since $a \neq 0$, we can divide by $2a$ to isolate b as $b = \frac{y}{2a}$ and substitute it into the first equation to get $x = a^2 - \frac{y^2}{4a^2}$ or

$$4a^4 - 4xa^2 - y^2 = 0.$$

By applying the quadratic formula to the "variable" a^2,

$$a^2 = \frac{x \pm \sqrt{x^2 + y^2}}{2}.$$

Recall from Theorem 9.13 that $|z| \geq |\text{Re}(z)|$, so $\sqrt{x^2 + y^2} \geq |x| \geq x$. This forces us to choose the positive sign in our expression from the quadratic formula, as otherwise a^2 would be non-positive. Then, since $a > 0$, we can take the positive square root of a^2 to get

$$a = \sqrt{\frac{\sqrt{x^2 + y^2} + x}{2}}.$$

Finally, by rationalizing the denominator of the expression shown below,

$$b = \frac{y}{2a} = \frac{y}{\sqrt{2}\sqrt{\sqrt{x^2 + y^2} + x}} = \frac{y}{\sqrt{y^2}} \cdot \sqrt{\frac{\sqrt{x^2 + y^2} - x}{2}}.$$

By Theorem 6.8,

$$\frac{y}{\sqrt{y^2}} = \frac{y}{|y|} = \text{sgn}(y)$$

and we do not have to consider the possibility of $y = 0$ because $z = x + iy$ is non-real. Therefore,

$$a + ib = \sqrt{\frac{\sqrt{x^2 + y^2} + x}{2}} + i \cdot \text{sgn}(y)\sqrt{\frac{\sqrt{x^2 + y^2} - x}{2}}.$$

The reader should algebraically verify that $(a + ib)^2 = x + iy$ is satisfied. □

Definition 10.13. Originally, we defined the coefficients of a quadratic function to be real constants. We can extend this to allow them to be complex constants. So we redefine a **quadratic function** to be a function $f : \mathbb{C} \to \mathbb{C}$ defined by $f(x) = ax^2 + bx + c$ for all complex numbers x, where a, b, c are complex constants. We are redefining the scope of quadratic functions now because Theorem 10.12 shows that every non-real complex number has exactly two complex square roots, just like non-zero reals; consequently, the quadratic formula and Vieta's formulas hold for quadratic functions with complex coefficients as well. The classification of solution types based on the sign of the discriminant does not extend as readily.

Problem 10.14. Suppose a, b, c are positive real numbers such that $a^2 - b^2c$ is non-negative. Show that the square root decomposition

$$\sqrt{a \pm b\sqrt{c}} = \sqrt{\frac{a + \sqrt{a^2 - b^2c}}{2}} \pm \sqrt{\frac{a - \sqrt{a^2 - b^2c}}{2}},$$

holds, where the $\pm$ signs on either side of the equation are both positive or both negative. In particular, if a, b, c are rational and $a^2 - b^2c$ is the square of a rational number, then we call this decomposition a "denesting" of square roots.

10.2 Graphing

Now, we will be assuming that the coefficients of the quadratic functions involved are real and that their domains are the real numbers. This will allow us to speak of the maximum or minimum possible output of a quadratic function, along with the location in the domain where this extreme value is attained.

Theorem 10.15. The quadratic function $f : \mathbb{R} \to \mathbb{R}$, defined by $f(x) = ax^2 + bx + c$ attains a global extremum of $-\frac{b^2 - 4ac}{4a}$ at $x = -\frac{b}{2a}$. By extremum, we mean that $f\left(-\frac{b}{2a}\right)$ is a minimum or maximum of $f(x)$ over all real x. This extremum is a minimum if $a > 0$ and a maximum if $a < 0$. When there is a global minimum, we say that the quadratic **opens up**, and, when there is is a global maximum, we say that the quadratic **opens down**.

Proof. Following the proof of Theorem 10.5, we complete the square, as follows:

$$\begin{aligned} f(x) &= ax^2 + bx + c \\ &= \frac{1}{4a}(4a^2x^2 + 4abx) + c \\ &= \frac{1}{4a}(4a^2x^2 + 4abx + b^2) - \frac{b^2}{4a} + c \\ &= \frac{1}{4a}(2ax + b)^2 - \frac{b^2 - 4ac}{4a} \\ &= a\left(x + \frac{b}{2a}\right)^2 - \frac{b^2 - 4ac}{4a}. \end{aligned}$$

Now we can use the trivial inequality. If $a > 0$, then

$$f(x) = a\left(x + \frac{b}{2a}\right)^2 - \frac{b^2 - 4ac}{4a} \geq -\frac{b^2 - 4ac}{4a},$$

where equality holds if and only if $x = -\frac{b}{2a}$. If $a < 0$, then

$$f(x) = a\left(x + \frac{b}{2a}\right)^2 - \frac{b^2 - 4ac}{4a} \leq -\frac{b^2 - 4ac}{4a},$$

where equality holds if and only if $x = -\frac{b}{2a}$ again. □

Graphs are defined in Volume 4. The graph curves formed by quadratic functions, called a parabola, are a part of a family of curves called conics, constructed by taking cross sections of cones. These are studied algebraically in Volume 4. Parabolas opening up or down are shown below.

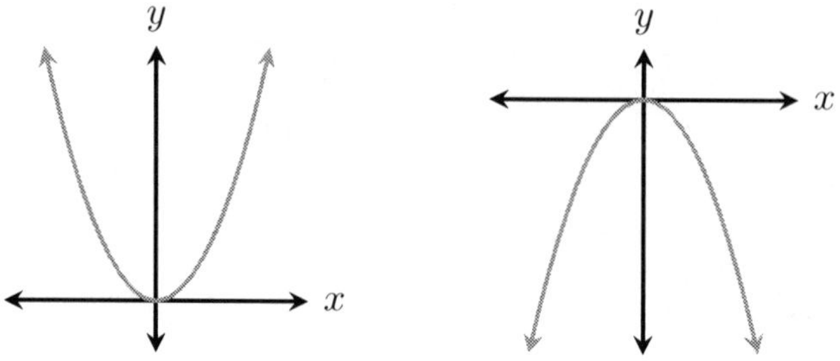

Definition 10.16. Since the quadratic $f(x) = ax^2 + bx + c$ attains a global extremum of $-\frac{b^2 - 4ac}{4a}$ at $x = -\frac{b}{2a}$, we call the point

$$\left(-\frac{b}{2a}, -\frac{b^2 - 4ac}{4a}\right)$$

on the quadratic's graph its **vertex**. It is for this reason that we call

$$a\left(x + \frac{b}{2a}\right)^2 - \frac{b^2 - 4ac}{4a}$$

the **vertex form** of f. The ordinary $ax^2 + bx + c$ form is called the **standard form**.

Corollary 10.17. Every quadratic function is a result of left-composing and right-composing the function $p(x) = x^2$ with linear functions.

Proof. The result is evident from the vertex form. To be formal, let $f(x) = ax^2 + bx + c$ be a quadratic function. We define the linear functions

$$g(x) = ax - \frac{b^2 - 4ac}{4a},$$
$$h(x) = x + \frac{b}{2a}.$$

Then $g \circ p \circ h = f$, thanks to the vertex form of f. As a result, the graph of f may be obtained by transforming the graph of the simple function $p(x) = x^2$ via linear deformations. □

Corollary 10.18. The graph of the quadratic function $f(x) = ax^2 + bx + c$ is symmetrical across the vertical line $x = -\frac{b}{2a}$ that runs through its vertex. This is the called the **line of symmetry** of the curve.

Proof. Writing f in vertex form and substituting in $x = -\frac{b}{2a} \pm r$ for any real r shows that

$$f\left(-\frac{b}{2a} + r\right) = ar^2 - \frac{b^2 - 4ac}{4a} = f\left(-\frac{b}{2a} - r\right).$$

This helps us to graph quadratics because the left or right half of the graph can be used to produce the other half. □

Chapter 11

Polynomials

"If I have seen further, it is by standing on the shoulders of giants."

– *Isaac Newton, letter to Robert Hooke*

Having seen constant, linear, and quadratic functions, we will now generalize them into polynomials. We will study the main parts of a polynomial, which are its degree, coefficients, and roots. Then we will take a look at the division of polynomials, which is the polynomial analogue of integer division with remainder. Thirdly, we will study rational functions, which are essentially one polynomial divided by another without remainder, but with domain restrictions that are necessary to avoid division by 0. Afterwards, we will look at a general version of Vieta's formulas, and a theorem of Girard and Newton. We will end the chapter by looking at factorizations of multivariable polynomials, including several well-known remarkable ones.

11.1 Degree, Coefficients, Roots

One way to motivate the definition of polynomials is to consider the functions or expressions that are achievable by multiplying and adding together several linear functions or expressions in the same variable. The general form that will be shown is how those functions would be presented if all factors were multiplied out and like terms were collected.

Definition 11.1. A **term** is an expression of the form ax^d, where a is a complex constant called its **coefficient**, x is the **variable** or **indeterminate**, and d is a non-negative integer called its **degree**. A univariate **polynomial** is a sum of a finite number of such terms, where terms of the same degree (called "like terms") are typically collected using the distributive law so that each term in the polynomial has a different degree. A polynomial takes the **standard form**

$$a_nx^n + a_{n-1}x^{n-1} + \cdots + a_2x^2 + a_1x + a_0.$$

Although complex numbers can be substituted in for the variable of a polynomial and the expression can then be evaluated using arithmetic operations, we caution the reader to not automatically interpret polynomials as functions. The definition of a **formal polynomial** is an infinite sequence of coefficients indexed by $\mathbb{Z}_+$,

$$(a_k)_{k=0}^{\infty} = (a_0, a_1, a_2, a_3, \ldots)$$

such that all elements of sufficiently high index are 0. We will make this idea more formal in the context of generating functions in Volume 2.

Definition 11.2. In a polynomial

$$f(x) = a_nx^n + a_{n-1}x^{n-1} + \cdots + a_2x^2 + a_1x + a_0,$$

the term a_nx^n with the highest degree is not allowed to have $a_n = 0$, and the degree n of this term is called the **degree** of the polynomial. It is denoted by $\deg f = n$. If all the coefficients of a polynomial are 0, then its degree is said to be $-\infty$, in order to be consistent with some relations involving degrees of polynomials that we will explore in Theorem 11.6, and we call that polynomial the **zero polynomial**. For ease of notation, if $\deg f = n$, then we define coefficients corresponding to terms with degree higher than n to be 0. That is, $a_k = 0$ for $k > \deg f$.

Definition 11.3. In a polynomial

$$f(x) = a_n x^n + a_{n-1} x^{n-1} + \cdots + a_2 x^2 + a_1 x + a_0,$$

we call $a_n \neq 0$ the **leading coefficient** and a_0 the **constant term**.

Example. A non-zero constant a is a polynomial of degree 0. A linear expression $ax + b$ is a polynomial of degree 1. A quadratic expression $ax^2 + bx + c$ is a polynomial of degree 2.

Definition 11.4. Two polynomials are said to be **equal** if they have the same degree and the coefficients of the two k^{th} degree terms are equal for all k, assuming all like terms have been collected. If we have two different expressions for the same polynomial (this often happens in algebra-based combinatorics), then we can set the coefficients of the corresponding terms equal to each other; this technique is called **comparing coefficients** and we will discuss it further in Volume 2. A separate, but related, concept is that of being **equal everywhere**, which means the two polynomials agree on the output value for every complex input (see Definition 11.8 for polynomials as functions). So being equal is in reference to formal polynomials, whereas being equal everywhere is in reference to polynomials as functions.

Definition 11.5. We define some specific binary and unary operations on formal polynomials in the following way so that the defining equations would be satisfied if any complex number were substituted in for x. Let there be two polynomials

$$f(x) = a_n x^n + a_{n-1} x^{n-1} + \cdots + a_2 x^2 + a_1 x + a_0,$$
$$g(x) = b_m x^m + b_{m-1} x^{m-1} + \cdots + b_2 x^2 + b_1 x + b_0.$$

1. **Addition** of polynomials is done term-by-term, so that

$$(f + g)(x) = \sum_{k=0}^{\max(n,m)} (a_k + b_k) x^k.$$

2. Taking the **negative** of a polynomial simply takes the negative of each coefficient so that

$$(-f)(x) = (-a_n) x^n + (-a_{n-1}) x^{n-1} + \cdots + (-a_2) x^2 + (-a_1) x + (-a_0).$$

   As such, we can define **subtraction** of polynomials as

$$f - g = f + (-g).$$

   Note that 0 is an additive identity of polynomials and $-f$ is an additive inverse of f.

3. **Multiplication** of polynomials is defined to be compatible with the arithmetic distributive law, so that

$$(f \cdot g)(x) = \sum_{k=0}^{n+m} c_k x^k,$$

   where

$$c_k = a_0 b_k + a_1 b_{k-1} + \cdots + a_{k-1} b_1 + a_k b_0.$$

4. **Composition** of polynomials is done by applying the distributive law and using the rules for adding and multiplying polynomials. It results in a polynomial, though we are unaware of a neat formula for the coefficients; interested readers may search for Faà di Bruno's formula. We use the ordinary composition notation $f \circ g$ to denote the composition of polynomials.

Theorem 11.6. For each type of operation on polynomials that we have described, we can comment on the degree of the resulting polynomials. If f and g are polynomials, then:

1. $\deg(-f) = \deg(f)$

2. $\deg(f \pm g) \leq \max(\deg f, \deg g)$, and, due to potential cancellation of terms, it is possible to achieve any of the values that can be degrees, which are $-\infty$ and non-negative integers. If $\deg g < \deg f$, we can be more precise and say $\deg(f \pm g) = \deg f$.

3. $\deg(f \cdot g) = \deg f + \deg g$

4. $\deg(f \circ g) = (\deg f) \cdot (\deg g)$ if f and g are non-constant polynomials

Proof. It is not difficult to see that the first result holds in general, the next two results hold for non-zero polynomials, and that the fourth result holds for non-constant polynomials. What we will discuss is how the second and third results force the convention that the degree of the 0 polynomial is $-\infty$. Let f be a non-zero polynomial and g be the 0 polynomial. For the product identity to be satisfied, it must be true that

$$\deg 0 = \deg(f \cdot g) = \deg f + \deg g = \deg f + \deg 0.$$

Since $\deg f$ could be any non-negative integer, there is no real value of $\deg 0$ that can satisfy this identity. However, if we make reasonable extensions of the rules of arithmetic to include $\pm\infty$ as "numbers," then these are candidates for the value of $\deg 0$. Now, in order for the addition identity to hold, we must have

$$\deg 0 = \deg(f - f) \leq \max(\deg f, \deg f) = \deg f.$$

This excludes the possibility of $\deg 0$ being $+\infty$, so the only candidate is $-\infty$. Indeed, it produces the following reasonable results for all polynomials f, including $f = 0$, under extended interpretations of arithmetic:

- $f = f + 0$ and $\deg f \leq \max(\deg f, -\infty) = \deg f$
- $0 = f - f$ and $-\infty \leq \max(\deg f, \deg(-f)) = \deg f$
- $0 = 0 \cdot f$ and $-\infty = -\infty + \deg f$

□

Problem 11.7. Prove that, if f is a non-constant polynomial then left-composing or right-composing f with a non-constant linear polynomial $ax + b$ leaves its degree unchanged.

Definition 11.8. A polynomial

$$f(x) = a_n x^n + a_{n-1}x^{n-1} + \cdots + a_2 x^2 + a_1 x + a_0$$

may be interpreted as a function $f : \mathbb{C} \to \mathbb{C}$ where we replace x with complex numbers z and evaluate

$$f(z) = a_n z^n + a_{n-1}z^{n-1} + \cdots + a_2 z^2 + a_1 z + a_0$$

using the rules of complex arithmetic. In compliance with the definition of a root of a general real- or complex-valued function (Definition 2.41), if z is a complex number such that $f(z) = 0$, then we call z a **root** or **solution** of f.

Problem 11.9. Show that, if all the coefficients of a polynomial are real and positive, then any real root must be negative.

Problem 11.10. Show that a polynomial has 0 as a root if and only if its constant term is 0.

Example 11.11. Let $f(x) = a_n x^n + a_{n-1}x^{n-1} + \cdots + a_2 x^2 + a_1 x + a_0$ be a polynomial where $a_0 \neq 0$. Show that z is a root of the polynomial with the reversed sequence of coefficients,

$$g(x) = a_0 x^n + a_1 x^{n-1} + \cdots + a_{n-2}x^2 + a_{n-1}x + a_n,$$

if and only if $\dfrac{1}{z}$ is a root of f.

Solution. The fact that f is of degree n lets us know that $a_n \neq 0$. Since $a_0 \neq 0$ and $a_n \neq 0$, the roots of f and g do not include 0, so we can take the reciprocal of any root of either polynomial. Suppose z is a root of g. Then

$$a_0 z^n + a_1 z^{n-1} + \cdots + a_{n-2} z^2 + a_{n-1} z + a_n = 0.$$

Dividing by z^n yields

$$a_0 + a_1 \left(\frac{1}{z}\right) + \cdots + a_{n-2}\left(\frac{1}{z}\right)^{n-2} + a_{n-1}\left(\frac{1}{z}\right)^{n-1} + a_n\left(\frac{1}{z}\right)^n = 0,$$

which shows that $\frac{1}{z}$ is a root of f. Since this step is reversible by multiplying by z^n, the other direction holds as well.

An interesting application of this result is to polynomials

$$f(x) = a_n x^n + a_{n-1}x^{n-1} + \cdots + a_2 x^2 + a_1 x + a_0$$

that are palindromic in the sense that $a_k = a_{n-k}$ for each index $0 \leq k \leq n$. Then z is a root of f if and only if $\frac{1}{z}$ is a root of f, which gives us a way of finding a new root from an existing root in this special case. We will see some other methods of finding new roots from old roots in Theorem 11.16 and Problem 11.18. As another side note, it may be readily verified that a polynomial f is palindromic if and only if it satisfies

$$f(x) = x^n \cdot f\left(\frac{1}{x}\right).$$

We will use this fact when studying cyclotomic polynomials in Volume 3. □

Theorem 11.12 (Rational root theorem)**.** If $\frac{p}{q}$ is a reduced fraction that is a rational root of a polynomial

$$f(x) = a_n x^n + a_{n-1}x^{n-1} + \cdots + a_2 x^2 + a_1 x + a_0$$

with all integer coefficients such that $a_0 \neq 0$, then p divides a_0 and q divides a_n. Since there are finitely many divisors of a_0 and a_n, we can iterate through all such rational numbers $\frac{p}{q}$ and test whether $f\left(\frac{p}{q}\right) = 0$ to find all the rational roots. (Note that p and q may be negative divisors of a_0 and a_n, respectively.)

Proof. Non-zero constant polynomials have no roots, so we can assume that $\deg f \geq 1$ so that $n \neq 0$. Suppose $\frac{p}{q}$ is a rational root of f. Then substituting $\frac{p}{q}$ into f yields

$$a_n\left(\frac{p}{q}\right)^n + a_{n-1}\left(\frac{p}{q}\right)^{n-1} + \cdots + a_2\left(\frac{p}{q}\right)^2 + a_1\left(\frac{p}{q}\right) + a_0 = 0.$$

We can clear all the denominators by multiplying both sides by q^n, which yields

$$a_n p^n + a_{n-1}p^{n-1}q + \cdots + a_2 p^2 q^{n-2} + a_1 p q^{n-1} + a_0 q^n = 0.$$

Since p divides the leftmost n terms on the left side, as well as the right side 0, p must also divide $a_0 q^n$. But we assumed that p is relatively prime to q, so p must divide a_0 by Gauss's divisibility lemma from Volume 3. Similarly, q divides the rightmost n terms on the left side, as well as the right side 0, so q must also divide $a_n p^n$. As q is relatively prime to p, it must be true that q divides a_n by Gauss's divisibility lemma. □

Corollary 11.13. Suppose

$$f(x) = a_n x^n + a_{n-1}x^{n-1} + \cdots + a_2 x^2 + a_1 x + a_0$$

is a non-zero polynomial. From the fact that f is non-zero, we know that there exists a non-negative integer k such that $a_k \neq 0$ with $a_{k-1} = \cdots = a_1 = a_0 = 0$ (if $k = 0$ then the list $a_{k-1}, \ldots, a_1, a_0$ is empty). Let

$$g(x) = a_n x^{n-k} + a_{n-1}x^{n-1-k} + \cdots + a_{k+2}x^2 + a_{k+1}x + a_k$$

be the polynomial such that $f(x) = x^k \cdot g(x)$. Then the non-zero roots of f are precisely the roots of g.

Proof. Let g be defined as $f(x) = x^k \cdot g(x)$ where x^k is the maximum power of x that can be factored out of f. If z is a non-zero root of f then

$$0 = f(z) = z^k g(z) \implies g(z) = 0.$$

And if z is a root of g then

$$f(z) = z^k g(z) = z^k \cdot 0 = 0.$$

As a consequence, even if the constant term of f is 0, we can still apply the rational root theorem to g to find all the rational roots of f. □

Definition 11.14. A **monic** polynomial is a polynomial whose leading coefficient is 1.

There are a further two important implications of the rational root theorem, both of which we label under the heading of the integer root theorem.

Corollary 11.15 (Integer root theorem). Let

$$f(x) = a_n x^n + a_{n-1} x^{n-1} + \cdots + a_2 x^2 + a_1 x + a_0$$

be a polynomial with integer coefficients and $a_n \neq 0$. Then:

1. Any non-zero integer root r of f divides a_k, where k is the smallest index i such that a_i is non-zero. As a consequence, r always divides a_0.
2. If f is monic, then all of its rational roots are integers.

Combining these two results, the only candidates for non-zero rational roots of a monic polynomial with integer coefficients are the factors of a_k, which is a finite list.

Proof. Let

$$f(x) = a_n x^n + a_{n-1} x^{n-1} + \cdots + a_2 x^2 + a_1 x + a_0$$

be a polynomial with integer coefficients. Let the polynomial g be defined by $f(x) = x^k \cdot g(x)$, where x^k is the maximum power of x that can be factored out of f. This means that $a_k \neq 0$ and

$$a_{k-1} = \cdots = a_1 = a_0 = 0.$$

Then:

1. Suppose r is a non-zero integer root of f. We know that r is a root of

$$g(x) = a_n x^{n-k} + a_{n-1} x^{n-1-k} + \cdots + a_{k+2} x^2 + a_{k+1} x + a_k.$$

   Since $a_k \neq 0$, we can apply the rational root theorem to the root $\frac{r}{1}$ and say that 1 divides the leading coefficient a_n (which is always true) and r divides a_k, as desired.

   If $a_0 = 0$ then r automatically divides a_0. If $a_0 \neq 0$, then $k = 0$ and r still divides $a_0 = a_k$.

2. Suppose f is monic, meaning $a_n = 1$. Let $r = \frac{p}{q}$ be a reduced rational root of f. If $r = 0$, then r is already an integer, so we can assume that r is a non-zero root. Then $\frac{p}{q}$ is a root of the polynomial g define above, and the rational root theorem says that p divides a_k and q divides 1. So $q = \pm 1$, which means $\frac{p}{q} = \pm p$ is an integer.

□

Notice that the quadratic formula tells us that, if a quadratic has two non-real roots, then the two roots are complex conjugates. This result generalizes, as follows.

Theorem 11.16. If f is a polynomial with real coefficients, and z is a complex root of f, then $\overline{z}$ is also a root of f.

Proof. This results follows from the fact that if z and w are complex numbers, then $\overline{z+w} = \overline{z} + \overline{w}$ and $\overline{z \cdot w} = \overline{z} \cdot \overline{w}$, and that if a is real then, $\overline{a} = a$. If

$$f(x) = a_n x^n + a_{n-1} x^{n-1} + \cdots + a_2 x^2 + a_1 x + a_0$$

is a polynomial with real coefficients and z is a complex root of f, then

$$\begin{aligned} f(\overline{z}) &= a_n (\overline{z})^n + a_{n-1} (\overline{z})^{n-1} + \cdots + a_2 (\overline{z})^2 + a_1 (\overline{z}) + a_0 \\ &= a_n \overline{z^n} + a_{n-1} \overline{z^{n-1}} + \cdots + a_2 \overline{z^2} + a_1 \overline{z} + a_0 \\ &= \overline{a_n z^n} + \overline{a_{n-1} z^{n-1}} + \cdots + \overline{a_2 z^2} + \overline{a_1 z} + \overline{a_0} \\ &= \overline{a_n z^n + a_{n-1} z^{n-1} + \cdots + a_2 z^2 + a_1 z + a_0} \\ &= \overline{f(z)} = \overline{0} = 0. \end{aligned}$$

So, $\overline{z}$ is also a root of f. □

Definition 11.17. Let a and b be rational numbers and c be a positive rational number that is not the square of a rational number. Then we say that the **radical conjugate** of $a + b\sqrt{c}$ is $\underline{a + b\sqrt{c}} = a - b\sqrt{c}$. We leave it to the reader to show that this definition makes sense because the representation $a + b\sqrt{c}$ is unique (meaning, if $a_1 + b_1\sqrt{c} = a_2 + b_2\sqrt{c}$, then $a_1 = a_2$ and $b_1 = b_2$.) This is not to be confused with the complex conjugate.

Problem 11.18. Let $r = a + b\sqrt{c}$ where a and b are rational numbers and c is a positive rational number that is not the square of a rational number. If f is a polynomial with rational coefficients and r is a root of f, then show that the radical conjugate $\underline{r}$ of r is also a root of f.

Theorem 11.19 (Equality of polynomials as functions)**.** Let

$$f(x) = a_n x^n + a_{n-1} x^{n-1} + \cdots + a_2 x^2 + a_1 x + a_0$$

be a polynomial with complex coefficients. Then the following are equivalent conditions on f.

1. All coefficients of f are 0.

2. For all complex z, we get $f(z) = 0$.

3. For $1 + \deg f$ many distinct complex z, we have $f(z) = 0$.

Consequently, the following conditions are equivalent on complex polynomials f and g:

a. The degrees of f and g are equal, and the coefficients of corresponding terms match.

b. For all complex z, we get $f(z) = g(z)$, meaning they are equal everywhere.

c. For $1 + \max(\deg f, \deg g)$ many distinct complex z, we have $f(z) = g(z)$.

Proof. For the first collection of equivalent statements, it is easy to see that (1) $\implies$ (2) $\implies$ (3), so it remains to be shown that (3) $\implies$ (1). Let $\deg f = n$, and let $z_0, z_1, z_2, \ldots, z_n$ be $n+1$ distinct complex numbers at which f evaluates to 0. This system of equations may be written in matrix form as

$$\begin{bmatrix} 1 & z_0 & z_0^2 & \cdots & z_0^{n-1} & z_0^n \\ 1 & z_1 & z_1^2 & \cdots & z_1^{n-1} & z_1^n \\ 1 & z_2 & z_2^2 & \cdots & z_2^{n-1} & z_2^n \\ \vdots & \vdots & \vdots & \ddots & \vdots & \vdots \\ 1 & z_{n-1} & z_{n-1}^2 & \cdots & z_{n-1}^{n-1} & z_{n-1}^n \\ 1 & z_n & z_n^2 & \cdots & z_n^{n-1} & z_n^n \end{bmatrix} \begin{bmatrix} a_0 \\ a_1 \\ a_2 \\ \vdots \\ a_{n-1} \\ a_n \end{bmatrix} = \begin{bmatrix} 0 \\ 0 \\ 0 \\ \vdots \\ 0 \\ 0 \end{bmatrix}.$$

The square matrix on the far left is a well-known matrix. It is called the Vandermonde matrix, and it has a non-zero determinant which can be beautifully computed. We will analyze it when studying special cases of the theory of linear recurrences in Volume 2. For now, what we need to know is that a non-zero determinant makes a matrix invertible, and multiplying the left sides of both sides of the equation by this inverse yields

$$(a_0, a_1, a_2, \ldots, a_{n-1}, a_n) = (0, 0, 0 \ldots, 0, 0).$$

For a separate proof that (3) $\implies$ (2), see Theorem 11.34.
For the second collection of equivalent statements, it is again easy to see that $(a) \implies (b) \implies (c)$. For $(c) \implies (a)$, we let $h = f - g$ so that $\deg h \leq \max(\deg f, \deg g)$ and $h(z) = 0$ for $1 + \max(\deg f, \deg g)$ many distinct complex z. By the earlier implication (3) $\implies$ (1), we get that all coefficients of $h = f - g$ are zero. This implies that $\deg f = \deg g$ and that the coefficients of f all match those of g. □

Example 11.20. Find an example of a ring or field in which there exists a polynomial that has some non-zero coefficients, but it is still equal to the additive identity, that is the "zero" field or ring element, for all inputs.

Solution. Let p be a prime. By Fermat's little theorem from Volume 3,

$$x^p - x \equiv 0 \pmod{p}$$

for every integer x. Therefore, it is possible for a polynomial with coefficients and inputs from $\mathbb{Z}_p$ to have non-zero coefficients while still being the identically "zero" function. □

11.2 Division and Factoring

Definition 11.21. So far, we have found results for univariate polynomials that have integer, rational, real, and complex coefficients. The sets of such formal polynomials are denoted by $\mathbb{Z}[x], \mathbb{Q}[x], \mathbb{R}[x]$, and $\mathbb{C}[x]$, respectively, and they are called **polynomial rings** over $\mathbb{S}$ where $\mathbb{S}$ is the set from which the coefficients are taken.

While we can define subtraction of polynomials easily, the polynomials rings $\mathbb{Z}[x]$, $\mathbb{Q}[x]$, $\mathbb{R}[x]$, and $\mathbb{C}[x]$ do not form fields because most polynomials do not have a multiplicative inverse that is a polynomial. As such, there is no way of always dividing a polynomial by another polynomial to get a polynomial. However, in some cases a polynomial does divide another polynomial, and in other cases we can do what is called Euclidean division, which is division with a remainder. In Section 11.3, we will look at rational functions, which are produced by dividing a polynomial function by a polynomial function without a remainder.

Definition 11.22. Let $\mathbb{S}$ denote one of the sets $\mathbb{Z}, \mathbb{Q}, \mathbb{R}$, or $\mathbb{C}$. If f and g are polynomials in $\mathbb{S}[x]$, then we say that f is **divisible** by g if there exists a polynomial $h \in \mathbb{S}[x]$ such that $f = g \cdot h$. Equivalently, we say that g is a **factor** of f or that g **divides** f.

Theorem 11.23 (Polynomial Euclidean division)**.** Let $\mathbb{F}$ be a field ($\mathbb{Q}, \mathbb{R}$, and $\mathbb{C}$ are examples of fields, but $\mathbb{Z}$ is not). For any $f \in \mathbb{F}[x]$ and any non-zero $g \in \mathbb{F}[x]$, there exists a $q \in \mathbb{F}[x]$ and an $r \in \mathbb{F}[x]$ such that

$$f = gq + r, \deg r < \deg g.$$

In this notation, if $\deg f \geq \deg g$ then $\deg q = \deg f - \deg g$. Moreover, given this f and non-zero g, the q and r are unique. We call f the **dividend**, g the **divisor**, q the **quotient**, and r the **remainder**. (By the same proof, the same result holds with every instance of $\mathbb{F}$ replaced with $\mathbb{Z}$ if g is monic. For a precise statement pertaining to the coefficients living in rings like $\mathbb{Z}$ instead of fields, read the proof of Theorem 11.54.)

Proof. Let $\mathbb{F}$ be a field, $f \in \mathbb{F}[x]$ and g be a non-zero element of $\mathbb{F}[x]$. First we will show that q and r exist, and then we will prove that they are unique. To prove existence, we will consider two cases:

1. If $\deg f < \deg g$, including the possibility that $f = 0$, then we can choose $q = 0$ and $r = f$ to get

$$f = g \cdot 0 + f = gq + r, \text{ and } \deg r < \deg g.$$

2. The more involved case is where $\deg f \geq \deg g$. We will proceed by strong induction on $\deg f$. Since $\deg g \geq 0$, the base case is $\deg f = 0$, meaning f is a non-zero constant a and g is a non-zero constant b. Since we must have $\deg r < \deg g$, the only choice for r is 0. Then $q = \frac{a}{b}$ and we get

$$f = a = b \cdot \frac{a}{b} + 0 = gq + r, \text{ and } \deg r < \deg g.$$

The strong induction hypothesis is that the assertion holds whenever $\deg f \leq n-1$ for some $n-1 \geq 0$. Now suppose $\deg f = n$. Let

$$f(x) = a_n x^n + a_{n-1}x^{n-1} + \cdots + a_2x^2 + a_1x + a_0,$$

and let $\deg g = m \leq n$ so that

$$g(x) = b_m x^m + b_{m-1}x^{m-1} + \cdots + b_2x^2 + b_1x + b_0.$$

If we could find a polynomial $h(x)$ such that $\deg(f - hg) < \deg f$, then either $f = hg + 0$ and we are done, or we can apply the induction hypothesis to $f - hg$ to get a quotient q_0 and a remainder r such that $f - hg = gq_0 + r$ with $\deg r < \deg g$. This would give us

$$f = g(h + q_0) + r,$$

so we could take $q = h + q_0$. To find h, we note that the leading term of hg must equal the leading term of f, so we let $h = \frac{a_n}{b_m}x^{n-m}$, which works. Notice that it is precisely in the division by b_m that we require the existence of multiplicative inverses in fields, which does not necessarily hold in arbitrary rings.

At this point, we compute can $\deg q$, now that we know that q and r exist. Since $\deg r < \deg g \leq \deg f$, we find that

$$\deg g + \deg q = \deg(g \cdot q) = \deg(f - r) = \deg f,$$

so $\deg q = \deg f - \deg g$.

Now we prove the uniqueness of q and r. Suppose $f = gq_1 + r_1$ with $\deg r_1 < \deg g$, and $f = gq_2 + r_2$ with $\deg r_2 < \deg g$. Then we get $gq_1 + r_1 = gq_2 + r_2$ or

$$g \cdot (q_1 - q_2) = r_2 - r_1.$$

For the sake of contradiction, suppose $q_1 - q_2 \neq 0$ so that $\deg(q_1 - q_2) \geq 0$. Then

$$\deg(r_2 - r_1) = \deg(g \cdot (q_1 - q_2)) = \deg g + \deg(q_1 - q_2) \geq \deg g.$$

However, we also know that

$$\deg(r_2 - r_1) \leq \max(\deg r_2, \deg r_1) < \deg g.$$

This is a contradiction so we must have $q_1 = q_2$, which implies that $r_1 = r_2$. □

The Euclidean division theorem for polynomials implicitly tells us that a quotient and a remainder exist, but it does not tell us how to find them. Sometimes we want to find them explicitly, which is where polynomial **long division** comes into play. Assuming that the reader is familiar with long division for integers, it will suffice to show an example of how polynomial long division works out. The idea is the same: if we are dividing f by g, then slice away at the terms of f from highest degree to lowest degree using multiples of g. Let us see an example.

Example 11.24. Find the quotient and remainder when $2x^3 + 11x + 6$ is divided by $x - 8$.

Solution. First we proceed by long division.

$$
\begin{array}{r|rrrrrrr}
 & 2x^2 & - & 16x & + & 117 & & \\
\hline
x+8 & 2x^3 & + & 0x^2 & - & 11x & + & 6 \\
 & -\ (2x^3 & + & 16x^2) & & & & \\
\cline{2-4}
 & & & -16x^2 & - & 11x & & \\
 & & - & (-16x^2 & - & 128x) & & \\
\cline{3-8}
 & & & & & 117x & + & 6 \\
 & & & & - & (117x & + & 936) \\
\cline{5-8}
 & & & & & & & -930
\end{array}
$$

The trouble with long division is that, well, it can take a long time to write out. Thankfully, there is a compressed way of writing down the same process when the divisor is a monic linear polynomial. It is called **synthetic division**. We write it out as follows.

$$
\begin{array}{r|rrrr}
-8 & 2 & 0 & -11 & 6 \\
 & & & & \\
\hline
 & & & & \multicolumn{1}{|r}{}
\end{array}
\quad \longrightarrow \quad
\begin{array}{r|rrrr}
-8 & 2 & 0 & -11 & 6 \\
 & & -16 & 128 & -936 \\
\hline
 & 2 & -16 & 117 & \multicolumn{1}{|r}{-930}
\end{array}
$$

So, the quotient is $2x^2 + 16x + 139$ and the remainder is 1118. The general process is:

1. Place the negative of the constant term of the divisor at the top-left corner.
2. Place the coefficients of the dividend in the top row, including any zero coefficients, sequenced from left to right in descending order of the degrees of the corresponding terms.
3. Each number in the second row is found by multiplying the number to its bottom-left by the number from the first step at the top-left corner.
4. Each number in the bottom row is the sum of the two numbers directly above it in the same column.

For example, the sequence in which we filled the entries above is: 2 in the bottom row, the -16 in the second row, the -16 in the bottom row, $128, 117, -936, -930$.
We caution the reader in several ways:

- As described, the above method only works when the divisor is a monic linear polynomial.
- If the divisor is $x - c$ then we place c in the top-left corner of the synthetic division box, not $-c$.
- If there are "missing" coefficients in the dividend, we still have to place them in the first line of the synthetic division box as 0.

As a bonus, synthetic division can be made to work for division by non-monic linear polynomials as well. If we are dividing f by $ax - b$ then first we use synthetic division on the problem of dividing f by $x - \dfrac{b}{a}$. Then

$$f(x) = \left(x - \frac{b}{a}\right) q(x) + r(x),$$

where q is the quotient and r is the remainder. This implies that

$$f(x) = (ax - b)\frac{q(x)}{a} + r(x),$$

so the true quotient is $\dfrac{q(x)}{a}$ and r is still the remainder. This follows from the fact that quotients and remainders are unique in the Euclidean division of polynomials. □

Problem 11.25. Find the quotient and remainder when $3x^4 + 10x + 5$ is divided by $2x^2 + x + 4$.

From here on, let $\mathbb{F}$ denote an arbitrary field, unless otherwise specified.

Theorem 11.26 (Polynomial remainder theorem). If $c \in \mathbb{F}$, then dividing $f \in \mathbb{F}[x]$ by $x - c$ yields a constant remainder equal to $f(c)$.

Proof. By Euclidean division,

$$f(x) = (x - c)q(x) + r(x),$$

with $\deg r < \deg(x - c) = 1$. So r must be a constant. Substituting $x = c$ into the equation yields

$$f(c) = r(c) = r.$$

We can generalize the argument to division by a general linear factor $ax - b$. Then the remainder would be $f\left(\frac{b}{a}\right)$, though the polynomial remainder theorem is usually stated only for division by monic linear polynomials. □

Theorem 11.27 (Polynomial factor theorem). If $c \in \mathbb{F}$, then $x - c$ divides $f \in \mathbb{F}[x]$ with zero remainder if and only if c is a root of f.

Proof. By the polynomial remainder theorem (Theorem 11.26), the remainder of f upon division by $x - c$ is $f(c)$. Thus, the polynomial factor theorem is equivalent to saying that $f(c) = 0$ if and only if c is a root of f, which follows from the definition of a root. □

Just like with integers in number theory, there a notion of a greatest common divisor of polynomials, along with related results, such as an analogue of Bézout's lemma for polynomials. While these are interesting, we will not discuss them. The reader who learns abstract algebra might come across these ideas in a more general setting.

Theorem 11.28 (Fundamental theorem of algebra). Every non-constant polynomial in $\mathbb{C}[x]$ has at least one complex root.

Proof. The proof of this theorem is beyond the scope of our material, as every known proof requires some form of higher mathematics. □

Definition 11.29. By the polynomial factor theorem (Theorem 11.27), $c \in \mathbb{F}$ is a root of $f \in \mathbb{F}[x]$ if and only if $x - c$ is a factor of f. We define the **multiplicity** of a root c to be the largest integer k such that $(x - c)^k$ is a factor of f.

Theorem 11.30. The following three conditions are equivalent for fields $\mathbb{F}$. For every non-constant polynomial $f \in \mathbb{F}[x]$:

1. f is the product $\deg f$ many monic linear polynomials from $\mathbb{F}[x]$, times a constant.
2. f has exactly $\deg f$ roots in $\mathbb{F}$, if the number of times that each root is counted is equal to its multiplicity. This is called counting "with multiplicity."
3. f has at least one root in $\mathbb{F}$.

Note that, while the fundamental theorem of algebra implies that they are all true for $\mathbb{F} = \mathbb{C}$, they might all fail for some other fields.

Proof. In a cyclic fashion, we will show that $1 \implies 2 \implies 3 \implies 1$.

- If $c \in \mathbb{F}$ such that $x - c$ is a factor of f then the polynomial factor theorem tells us that c is a root of f (Theorem 11.27). Since we are assuming that f has $\deg f$ monic linear factors $x - c$, each c is a root. Counting roots with multiplicity, this means there are at *least* $\deg f$ roots of f in $\mathbb{F}$.

  Now we will show by induction on $\deg f$ that there are at *most* $\deg f$ roots of f in $\mathbb{F}$. We need to prove the result for $\deg f \geq 1$, but for the sake of convenience, we will start with the base case $\deg f = 0$, where f is a non-zero constant. If f is a non-zero constant, then it has no roots, and so it is true that f has at most $\deg f$ roots.

For the induction hypothesis, suppose that the result holds for all non-zero polynomials of degree less than or equal to n for some integer $n \geq 0$, and let f be a polynomial of degree $n+1$. If f has no roots, then it is true that f has at most $\deg f$ roots. If f has a root c then let k be the multiplicity of c so that

$$f(x) = (x-c)^k q(x)$$

for some $q \in \mathbb{F}[x]$. The polynomial factor theorem tells us that $k \geq 1$, so

$$n+1 = \deg f = \deg((x-c)^k) + \deg q = k + \deg q \geq 1 + \deg q.$$

Then $\deg q \leq n+1-1 = n$. By the induction hypothesis, q has at most $n+1-k$ roots counted with multiplicity, none of which are c. Moreover, every root of f is either c, or, if it is $d \neq c$, then d is a root of q because

$$0 = f(d) = (d-c)^k q(d) \implies q(d) = 0.$$

Therefore, f has at most

$$k + (n+1-k) = n+1 = \deg f$$

roots counted with multiplicity, as desired.

Since we have shown that f has both at least and at most $\deg f$ roots with multiplicity, f has exactly $\deg f$ roots with multiplicity, by antisymmetry.

- If f has exactly $\deg f \geq 1$ roots in $\mathbb{F}$, then it must have at least one root in $\mathbb{F}$.
- Suppose f has at least one root in $\mathbb{F}$. We will prove by induction on $\deg f \geq 1$ that f is the product of $\deg f$ many monic polynomials that are all from $\mathbb{F}[x]$, times a constant. In the base case, $\deg f = 1$ and so must have the form $ax + b = a\left(x + \frac{b}{a}\right)$. So the result holds for linear polynomials.

  For the induction hypothesis, we will assume that the result holds for polynomials with degree equal to n for some $n \geq 1$. Suppose f has degree $n+1$. Since we know that f has a root c, the polynomial factor theorem tells us that $f(x) = (x-c)q(x)$ for some $q \in \mathbb{F}[x]$. We know that

$$\deg f = \deg(x-c) + \deg q = 1 + \deg q,$$

  so $\deg q = \deg f - 1 < \deg f$, which allows us to apply the induction hypothesis on q. This tells us that q is a product of a constant times $\deg q = (\deg f) - 1$ monic linear factors. The rest follows because we just tack on an extra factor $x - c$, according to $f(x) = (x-c)q(x)$.

As a side note, since f has exactly $\deg f$ roots with multiplicity, f has at most $\deg f$ *distinct* roots. □

Problem 11.31. Prove that, if p and q are polynomials in $\mathbb{C}[x]$ such that pq is the zero polynomial, then at least one of p or q is the zero polynomial. A consequence of this is that, if f, g, h are polynomials such that $fg = fh$ and f is not the zero polynomial, then $f \cdot (g-h) = 0$ and we can "cancel" f from both sides to get $g = h$.

Corollary 11.32. If f is a non-constant polynomial in $\mathbb{C}[x]$ of degree n, then it has exactly n complex roots $z_1, z_2, \ldots, z_n$, including multiplicity, and f factors as

$$a(x-z_1)(x-z_2)\cdots(x-z_n),$$

where a is its leading coefficient, and the factorization is unique up to reordering the factors $x - z_j$. As a result, if two polynomials $f, g \in \mathbb{C}[x]$ have the same set of roots (and so their degrees are equal), then $f = \alpha g$ for some non-zero $\alpha \in \mathbb{R}$; if we further know that the leading coefficients of f, g are equal, then $f = g$.

Proof. We know from Theorem 11.30 that f has exactly n roots $z_1, z_2, \ldots, z_n$ with multiplicity and we know that one way of factoring f is

$$f(x) = a(x - z_1)(x - z_2) \cdots (x - z_n),$$

where a is its leading coefficient. So we just need to prove uniqueness. Suppose there is another factorization

$$f(x) = b(x - w_1)(x - w_2) \cdots (x - w_m).$$

Then we can equate them to get

$$a(x - z_1)(x - z_2) \cdots (x - z_n) = b(x - w_1)(x - w_2) \cdots (x - w_m).$$

The leading terms must be equal, meaning $ax^n = bx^m$, which implies $n = m$ and subsequently $a = b$. Substituting w_1 into both sides yields

$$(w_1 - z_1)(w_1 - z_2) \cdots (w_1 - z_n) = 0,$$

which means $w_1 = z_i$ for some $i \in [n]$. Reordering the z_j if necessary, we may assume without loss of generality that $i = 1$. Via polynomial division, we can remove $x - z_1 = x - w_1$ from both sides to get

$$(x - z_2) \cdots (x - z_n) = (x - w_2) \cdots (x - w_n).$$

Continuing in this way n times, we find that the multisets (see Volume 2)

$$\langle z_1, z_2, \ldots, z_n \rangle \text{ and } \langle w_1, w_2, \ldots, w_m \rangle$$

are identical. So, the factorization is unique up to reordering the factors.
For the consequences, if f and g have the same roots $z_1, z_2, \ldots, z_n$, then there exist non-zero $a, b \in \mathbb{R}$ such that

$$f(x) = a(x - z_1)(x - z_2) \cdots (x - z_n),$$
$$g(x) = b(x - z_1)(x - z_2) \cdots (x - z_n),$$

so f, g differ by only a constant factor. Of course, if $a = b$, then $f = g$. □

Theorem 11.33. The list of complex roots of a non-constant polynomial $f \in \mathbb{R}[x]$ is

$$r_1, r_2, \ldots, r_i, z_1, \overline{z_1}, z_2, \overline{z_2}, \ldots, z_j, \overline{z_j}$$

for non-negative integers i and j, where the r_k are real and the z_k and $\overline{z_k}$ are non-real complex numbers that come in such conjugate pairs, and $i + 2j = \deg f$. As a consequence, the number of non-real complex roots of f is even.

Proof. Let f be a non-constant polynomial of degree n with real coefficients, and let its list of roots be $w_1, w_2, \ldots, w_n$. Then

$$f(x) = a(x - w_1)(x - w_2) \cdots (x - w_n),$$

where a is its leading coefficient. If all the w_k are real, then there nothing left to prove. Otherwise, suppose a non-real complex real root exists. Since the w_k may be reordered without changing the factored expression for f, we can assume without loss of generality that w_1 is a non-real complex root. We know from Theorem 11.16 that $\overline{w_1}$ is also a root, so we also assume without loss of generality that $w_2 = \overline{w_1}$. Then the product of the first two factors is

$$(x - w_1)(x - w_2) = (x - w_1)(x - \overline{w_1}) = x^2 - (w_1 + \overline{w_1})x + w_1\overline{w_1} = x^2 - 2\text{Re}(w_1) + |w_1|^2,$$

which is a polynomial with real coefficients. The Euclidean division theorem then tells us that the quotient

$$(x - w_3)(x - w_4) \cdots (x - w_n)$$

after dividing f by $(x - w_1)(x - w_2)$ also has real coefficients. This allows us to continue repeating the process by taking away pairs of conjugate non-real complex roots until we run out of non-real complex roots. □

Theorem 11.34 (Polynomial identity theorem). The fundamental theorem of algebra and its corollaries help us to derive the more difficult parts of Theorem 11.19 in a new way, as follows.

1. If $f \in \mathbb{C}[x]$ and f has strictly more than $\deg f$ distinct complex roots, then f is zero everywhere.
2. Suppose f and g are polynomials in $\mathbb{C}[x]$ such that both have degree less than or equal to n. If $f(z) = g(z)$ for more than n distinct points $z \in \mathbb{C}$, then f and g are equal everywhere.

These results help us to identify polynomials using a bounded number of points.

Proof. The latter follows from the former, so we will prove them in sequence:

1. One consequence (Theorem 11.30) of the fundamental theorem of algebra is that if $\deg f \geq 1$ then f has at most $\deg f$ distinct roots. By contrapositive, if f has more than $\deg f$ distinct roots, then f is a constant polynomial. If f is a non-zero constant, then it has no roots, which contradicts the requirement that it must have strictly more than $\deg f = 0$ roots. So, instead, f must be zero everywhere.
2. If $f(z) = g(z)$ for more than n complex numbers z, then $(f - g)(z) = 0$ at each of those points. But we know that
$$\deg(f - g) \leq \max(\deg f, \deg g) \leq n,$$
so the fundamental theorem of algebra tells us that $f - g$ is zero everywhere as in the preceding part. This implies that f and g are equal everywhere.

□

11.3 Rational Functions

Definition 11.35. A **rational function** is a function that can be expressed in the form $f(x) = \dfrac{p(x)}{q(x)}$, where p and q are polynomial functions, and q is not the zero polynomial. For our purposes, they will only be polynomials with real coefficients. The domain of f is restricted to those real numbers that are not real roots of q, in order to avoid division by 0.

Theorem 11.36. If a rational function f can be written in the form $\dfrac{p(x)}{q(x)}$, where p and q are polynomials and q is not the zero polynomial, then the real roots of f are the real roots of p, excluding any real roots of q.

Proof. Suppose $f(r) = 0$ for some real r at which f is defined. Then $q(r) \neq 0$, otherwise f would not be defined at r. And $f(r) = \dfrac{p(r)}{q(r)} = 0$ implies $p(r) = 0 \cdot q(r) = 0$, so r is a real root of p, Conversely, if r is a real root of p but not a real root of q, then f is defined at r and
$$f(r) = \frac{p(r)}{q(r)} = \frac{0}{q(r)} = 0.$$
So, r is also a real root of f. □

Definition 11.37. An **asymptote** of a function f is a line to which the f gets arbitrarily close as the inputs of the function approach a certain real value or $+\infty$ or $-\infty$. For our purposes, it is acceptable for f to intersect an asymptote, even infinitely many times. We can classify the notion of asymptote into three types of lines: vertical, horizontal, or oblique, the last of which is also called slant.

1. A **vertical asymptote** is a line of the form $x = c$ for a real constant c. This line may be an asymptote, regardless of whether f is defined at c, or what value $f(c)$ takes on if f does happen to be defined at c. The asymptotic behaviour of f near c comes in four flavours:

- $f(x)$ approaches $-\infty$ from the left of $x = c$ but $+\infty$ from the right of $x = c$
- $f(x)$ approaches $+\infty$ from the left of $x = c$ but $-\infty$ from the right of $x = c$
- $f(x)$ approaches $+\infty$ from both the left and right of $x = c$
- $f(x)$ approaches $-\infty$ from both the left and right of $x = c$

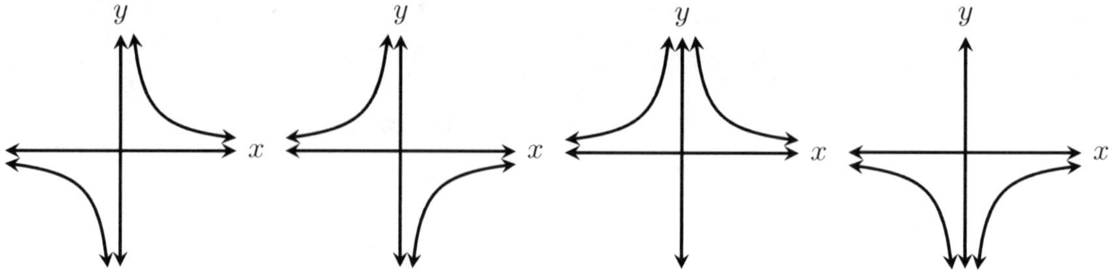

2. A **horizontal asymptote** is a line of the form $y = c$ that $f(x)$ approaches as $x \to \infty$ or $x \to -\infty$.

3. An **oblique or slant asymptote** is a line $g(x) = ax + b$ with $a \neq 0$ such that $f(x) - g(x)$ goes to 0 as $x \to \infty$ or $x \to -\infty$.

For an arbitrary function, there can be different horizontal or oblique asymptotes in the two directions.

Theorem 11.38. We can classify when asymptotes occur for rational functions and develop methods of finding them. Let $f(x) = \dfrac{p(x)}{q(x)}$ be a rational function, where p and q are polynomials. Then:

1. f has a vertical asymptote at $x = c$ if and only if the multiplicity of $x - c$ in q is greater than the multiplicity of $x - c$ in p.

2. f has a horizontal asymptote as $x \to \infty$ if and only if it has a horizontal asymptote as $x \to -\infty$. In this case, the two lines are equal, and we call this line the horizontal asymptote of f. Moreover, f has a horizontal asymptote if and only if $\deg p \leq \deg q$. If $\deg p < \deg q$ then the horizontal asymptote is $y = 0$. If $\deg p = \deg q$ then the horizontal asymptote is $y = \dfrac{a}{b}$, where a is the leading coefficient of p and b is the leading coefficient of q.

3. As with horizontal asymptotes, f has an oblique asymptote as $x \to \infty$ if and only if it has an oblique asymptote as $x \to -\infty$. In this case, the two lines are equal, and we call this line the oblique asymptote of f. Moreover, f has an oblique asymptote if and only if $\deg p = 1 + \deg q$. In this case, the oblique asymptote is the quotient found upon doing Euclidean division of p by q.

A rational function f can have an arbitrarily large finite number of vertical asymptotes, but it is restricted to at most one horizontal or oblique asymptote. That is, the existence of a horizontal asymptote excludes the existence of an oblique asymptote, and vice versa.

Proof. Since asymptotes involve the notion of "approaching," formal proofs of these assertions would be done in the language of limits from calculus. As this is beyond the scope of our exposition, we will provide only intuitive arguments for the "if" direction of the classification of when each type of asymptote occurs, along with informal justification for the technique used to compute the asymptote. Let f be as defined in the problem.

1. If $q(c) = 0$ and $p(c) \neq 0$, then we are essentially "dividing by 0" in $\dfrac{p(x)}{q(x)}$, so the behaviour of f around c is to shoot up to ∞ or down to ∞.

2. Suppose $m = \deg p \leq \deg q = n$. Let

$$p(x) = a_m x^m + a_{m-1} x^{m-1} + \cdots + a_2 x^2 + a_1 x + a_0,$$
$$q(x) = b_n x^n + b_{n-1} x^{n-1} + \cdots + b_2 x^2 + b_1 x + b_0.$$

The general idea is to divide the numerator and denominator of $\dfrac{p(x)}{q(x)}$ by x^n and observe what happens to each term of p and q as $x \to \infty$. This gives

$$\frac{p(x)}{q(x)} = \frac{\frac{a_m}{x^{n-m}} + \frac{a_{m-1}}{x^{n-m+1}} + \cdots + \frac{a_2}{x^{n-2}} + \frac{a_1}{x^{n-1}} + \frac{a_0}{x^n}}{b_n + \frac{b_{n-1}}{x} + \cdots + \frac{b_2}{x^{n-2}} + \frac{b_1}{x^{n-1}} + \frac{b_0}{x^n}}.$$

If $m < n$ then every term in the numerator and denominator, except b_n in the denominator, goes to 0 as $x \to \infty$. So $\dfrac{p(x)}{q(x)} \to \dfrac{0}{b_n}$ meaning $f(x) \to 0$ as $x \to \infty$. If $m = n$, then the numerator goes to a_m and the denominator goes to b_n as $x \to \infty$, so $f(x) \to \dfrac{a_m}{b_n}$.

3. Suppose $\deg p = 1 + \deg q$. By the Euclidean division theorem for polynomials,

$$p = qs + r$$

for a quotient s and a denominator r, where $\deg r < \deg q$ and $\deg s = \deg q - \deg p = 1$. Then

$$\frac{p}{q} - s = \frac{r}{q},$$

so the behaviour of $f - s$ the same as the behaviour of $\dfrac{r}{q}$ as $x \to \infty$. But $\deg r < \deg q$, so by the same argument as for horizontal asymptotes, $\dfrac{r}{q} \to 0$ as $x \to \infty$. Thus, s is an oblique asymptote of f.

□

Now that we have explored roots and asymptotes of rational functions, we are ready to tackle the question of when a rational function takes on what sign. We will need the following lemma as a prerequisite.

Lemma 11.39. Every polynomial $f \in \mathbb{R}[x]$ can be written as a product of a constant, monic linear polynomials $x + a$ where $a \in \mathbb{R}$, and monic quadratic polynomials $x^2 + bx + c$ where $b, c \in \mathbb{R}$ such that $b^2 - 4c < 0$. The linear or quadratic components might be empty.

Proof. If f is a constant, the result is immediate. If f is non-constant, then Theorem 11.33 says that its list of roots is

$$r_1, r_2, \ldots, r_i, z_1, \overline{z_1}, z_2, \overline{z_2}, \ldots, z_j, \overline{z_j},$$

where the r_k are real and the z_k and $\overline{z_k}$ are non-real complex numbers, and $i + 2j = \deg f$. Then

$$f(x) = (x - r_1) \cdots (x - r_i)(x - z_1)(x - \overline{z_1}) \cdots (x - z_j)(x - \overline{z_j}).$$

We can choose the monic linear factors to be $x - r_k$ and the monic quadratic factors to be $(x - z_k)(x - \overline{z_k})$. This works because

$$(x - z_k)(x - \overline{z_k}) = x^2 - 2\text{Re}(z_k) + |z_k|^2,$$

where the coefficients are real and

$$|z_k|^2 = (\text{Re}(z_k))^2 + (\text{Im}(z_k))^2 > (\text{Re}(z_k))^2,$$

so the discriminant is

$$(2\text{Re}(z_k))^2 - 4 \cdot 1 \cdot |z_k|^2 < (2\text{Re}(z_k))^2 - 4 \cdot (\text{Re}(z_k))^2 = 0.$$

In other words, each quadratic factor cannot split into two linear factors with real coefficients. □

Let us now see the process of solving a rational inequality that contains every possible obstacle. The method will work for polynomials too, as the denominator is then simply 1.

Example 11.40 (Interval analysis). Find all real x such that

$$f(x) = \frac{-6(x-2)(x+1)^2(x+3)^3(x^2+x+2)}{7(x-2)(x-4)(x+1)(x+5)^4} \geq 0.$$

Solution. First we split this into two problems: finding the real roots of f and finding when the strict inequality $f(x) > 0$ holds. The real roots of f are the real roots of the numerator, excluding the real roots of the denominator. This gives the set

$$\{2, -1, -3\} \backslash \{2, 4, -1, -5\} = \{-3\}.$$

Now we will find when the inequality holds strictly and pop -3 back into the solution in the end. We take f over to the other side to remove the pesky negative sign, and then we remove the leading coefficients to get the equivalent inequality

$$\frac{(x-2)(x+1)^2(x+3)^3(x^2+x+2)}{(x-2)(x-4)(x+1)(x+5)^4} < 0, x \neq 2, 4, -1, -5.$$

Then we cancel out pairs of common linear factors between the numerator and denominator, while noting that the domain of the simplified expression cannot include the roots of those cancelled factors. This gives the equivalent inequality

$$\frac{(x+1)(x+3)^3(x^2+x+2)}{(x-4)(x+5)^4} < 0, x \neq 2, 4, -1, -5.$$

Since the quadratic factor's discriminant is negative and its leading coefficient is positive, it always takes on positive values, so we can divide out by it to get the equivalent inequality

$$\frac{(x+1)(x+3)^3}{(x-4)(x+5)^4} < 0, x \neq 2, 4, -1, -5.$$

The squares of linear factors are always non-negative, so $(x+3)^2 \geq 0$ and $(x+5)^4 \geq 0$. Moreover, it is not possible that $x = -3$, since that would make the whole expression 0, contradicting the fact that we are working on a strict inequality. It is also not possible that $x = -5$, since that is not in the domain of the expression. So $(x+3)^2 > 0$ and $(x+5)^4 > 0$ and we can divide out by them to get the equivalent inequality

$$\frac{(x+1)(x+3)}{x-4} < 0, x \neq 2, 4, -1, -5.$$

Now the expression is simple enough for us to perform **interval analysis**. The list of roots is $-1, -3$ and the list of singularities (i.e. where vertical asymptotes occur) is 4. We will call the list $-3, -1, 4$ the list of "transition points," which is not standard terminology. It should intuitively make sense that it is only at roots and singularities that the sign might change. So, the function that the expression $\frac{(x+1)(x+3)}{x-4}$ represents always has the same sign when x is strictly in between two consecutive change points. This idea allows us to partition the real line into intervals and test the sign of each factor in each interval.

	$(-\infty, -3)$	$(-3, -1)$	$(-1, 4)$	$(4, \infty)$
$x+1$	$-$	$-$	$+$	$+$
$x+3$	$-$	$+$	$+$	$+$
$x-4$	$-$	$-$	$-$	$+$
$f(x)$	$-$	$+$	$-$	$+$

So our preliminary solution is $(-\infty, -3) \cup (-1, 4)$. We then remove the points $\{2, 4, -1, -5\}$ because they are not in the domain of f, which refines the solution to

$$(-\infty, -5) \cup (-5, -3) \cup (-1, 2) \cup (2, 4).$$

Finally, we cannot forget to include the root -3, which gives the final solution

$$(-\infty, -5) \cup (-5, -3] \cup (-1, 2) \cup (2, 4).$$

Those who are interested in knowing why interval analysis works should look into the intermediate value theorem from calculus. □

A general algorithm for solving rational inequalities should be evident from the example provided. It is essentially a matter of going through a sequence of simplifications, while ensuring that the steps are reversible, and then analyzing the signs of the remaining factors on each open interval defined by consecutive change points. The only step that we did not show is that of factoring the numerator and denominator into linear and irreducible quadratic factors. The fact that this is possible is proven by Lemma 11.39, and we have developed several techniques to make it happen if the circumstances are sufficiently convenient: Theorem 11.12, Corollary 11.15, Theorem 11.16, and Problem 11.18. Even Example 11.11 is helpful in special cases. Another trick is that z is a root if and only if $-z$ is a root, if all terms of odd degree have 0 as the coefficient, since this leaves only even degree terms.

11.4 Symmetry

Definition 11.41. A **term** ax^α in n variables $x_1, x_2, \ldots, x_n$ is an expression of the form $ax_1^{\alpha_1}x_2^{\alpha_2}\cdots x_n^{\alpha_n}$, where the **coefficient** a is an element of a fixed field $\mathbb{F}$, such as $\mathbb{C}$, and the α_i are non-negative integers. The **multidegree** of this term is the multi-index $\alpha = (\alpha_1, \alpha_2, \ldots, \alpha_n)$ and its **degree** is $|\alpha| = \alpha_1+\alpha_2+\cdots+\alpha_n$. A **polynomial** in the n variables $x_1, x_2, \ldots, x_n$ is a sum of finitely many terms in these variables. The **degree** of a polynomial is the highest degree among the degrees of its non-zero terms. The set of polynomials in these n distinct variables with coefficients in the field $\mathbb{F}$ is denoted by $\mathbb{F}[x_1, x_2, \ldots, x_n]$.

Problem 11.42. By the quadratic variant of Vieta's formulas (Corollary 10.10), if the roots of $ax^2 + bx + c$ are r_1 and r_2 then

$$\begin{aligned} r_1 + r_2 &= -\frac{b}{a} \\ r_2 r_2 &= \frac{c}{a}. \end{aligned}$$

Find analogous formulas for cubic equations, which are univariate polynomials of degree 3.

Inspired by these results for quadratic and cubic polynomials, we will generalize them to non-constant polynomials.

Theorem 11.43 (Vieta's formulas)**.** Suppose

$$f(x) = a_n x^n + a_{n-1}x^{n-1} + \cdots + a_2x^2 + a_1x + a_0$$

is a polynomial of degree $n \geq 1$ with complex coefficients a_i. Then the complex $r_1, r_2, \ldots, r_n$ are the n roots of f if and only if the following n equations hold:

$$\begin{aligned} r_1 + r_2 + \cdots + r_n &= (-1)^1 \cdot \frac{a_{n-1}}{a_n}, \\ (r_1r_2 + \cdots + r_1r_n) + \cdots + (r_{n-1}r_n) &= (-1)^2 \cdot \frac{a_{n-2}}{a_n}, \\ &\vdots \\ \sum_{\substack{J\subseteq[n] \\ |J|=k}} \prod_{j\in J} r_j = \sum_{1\leq i_1<i_2<\cdots<i_k\leq n} r_{i_1}r_{i_2}\cdots r_{i_k} &= (-1)^k \cdot \frac{a_{n-k}}{a_n}, \\ &\vdots \\ r_1r_2\cdots r_n &= (-1)^n \cdot \frac{a_0}{a_n}. \end{aligned}$$

Proof. For the main direction, suppose $r_1, r_2, \ldots, r_n$ are the n roots of f. We could perform a dry proof by induction on n, but it will be more instructive to consider the expansion of

$$\begin{aligned} f(x) &= a_n x^n + a_{n-1}x^{n-1} + \cdots + a_2x^2 + a_1x + a_0 \\ &= a_n(x-r_1)(x-r_2)\cdots(x-r_n). \end{aligned}$$

Dividing both sides by a_n yields

$$x^n + \frac{a_{n-1}}{a_n}x^{n-1} + \cdots + \frac{a_2}{a_n}x^2 + \frac{a_1}{a_n}x + \frac{a_0}{a_n} = (x-r_1)(x-r_2)\cdots(x-r_n).$$

Now we will discover the coefficients of the right side upon expansion. In the expansion, before like terms are collected, there are 2^n terms, each having been created by choosing x from $n-k$ of the factors $x - r_i$ and $-r_i$ from the remaining k factors. If we want the coefficient of x^{n-k} on the right side, we will choose x from $n-k$ of the factors and $-r_i$ from the remaining k factors. This yields exactly all of the terms of the form

$$(-r_{i_1})(-r_{i_2})\cdots(-r_{i_k})x^{n-k} = (-1)^k \cdot r_{i_1}r_{i_2}\cdots r_{i_k}x^{n-k}$$

for all integer multi-indices $(i_1, i_2, \ldots, i_k)$ such that $1 \le i_1 < i_2 < \cdots < i_k \le n$. But the coefficient of x^{n-k} on the left side is $\frac{a_{n-k}}{a_n}$, so

$$\sum_{\substack{J \subseteq [n] \\ |J|=k}} \prod_{j \in J} r_j = \sum_{1 \le i_1 < i_2 < \cdots < i_k \le n} r_{i_1}r_{i_2}\cdots r_{i_k} = (-1)^k \cdot \frac{a_{n-k}}{a_n},$$

for each integer k such that $1 \le k \le n$, as expected.

Conversely, suppose $r_1, r_2, \ldots, r_n$, are n complex numbers such that the n stated equations hold. According to our described expansion,

$$\begin{aligned} a_n(x-r_1)(x-r_2)\cdots(x-r_n) &= a_nx^n + \sum_{k=1}^{n}(-1)^k a_n \left[\sum_{\substack{J \subseteq [n] \\ |J|=k}} \prod_{j \in J} r_j\right] x^{n-k} \\ &= a_nx^n + \sum_{k=1}^{n} a_{n-k}x^{n-k} = f(x). \end{aligned}$$

According to Corollary 11.32, f has a unique factorization into linear factors, so $r_1, r_2, \ldots, r_n$ are the n roots of f. □

Definition 11.44. Let n, k be positive integers such that $k \neq n$. Given n complex numbers $r_1, r_2, \ldots, r_n$, their k^{th} **symmetric sum** is defined as

$$\sigma_k = \sum_{\substack{J \subseteq [n] \\ |J|=k}} \prod_{j \in J} r_j = \sum_{1 \le i_1 < i_2 < \cdots < i_k \le n} r_{i_1}r_{i_2}\cdots r_{i_k}.$$

Corollary 11.45. Since factoring and expanding are two sides of the same coin that is the distributive law, we can reverse-engineer the roots in Vieta's formulas. That is, suppose we are given the n symmetric sums $\sigma_1, \sigma_2, \ldots, \sigma_n$ of n unknown complex numbers $r_1, r_2, \ldots, r_n$. Then a monic polynomial of degree n in $\mathbb{C}[x]$ with $r_1, r_2, \ldots, r_n$ as its roots is

$$(x-r_1)(x-r_2)\cdots(x-r_n) = x^n + \sum_{k=1}^{n}(-1)^k \sigma_k x^{n-k}.$$

The right side is known to us because we know the value of each σ_k, and subsequently we can often obtain information about the r_i using various techniques available to us, such as the rational root theorem (Theorem 11.12), integer root theorem (Corollary 11.15), and the fact that non-real complex roots come in conjugate pairs (Theorem 11.16).

Definition 11.46. A polynomial $f \in \mathbb{F}[x_1, x_2, \ldots, x_n]$ is said to be **symmetric** if swapping any two variables x_i and x_j in f results in the same polynomial.

Example. The polynomial
$$f(x, y) = x^3 + xy^2 + x^2y + y^3$$
is symmetric because
$$f(y, x) = y^3 + yx^2 + y^2x + x^3 = f(x, y),$$
thanks to commutativity of addition and multiplication. We swapped every instance of x with y and vice versa, and checked whether this produces an equivalent polynomial.

Definition 11.47. Let n, k be positive integers such that $k \neq n$. Given n variables $x_1, x_2, \ldots, x_n$, the kth **elementary symmetric polynomial** in $\mathbb{F}[x_1, x_2, \ldots, x_n]$ is defined as
$$e_k(x_1, x_2, \ldots, x_n) = \sum_{1 \le i_1 < i_2 < \cdots < i_k \le n} x_{i_1} x_{i_2} \cdots x_{i_k}.$$

Lemma 11.48. Let n be a positive integer. Every polynomial in $\mathbb{F}[x_1, x_2, \ldots, x_n, x_{n+1}]$ may be interpreted as a univariate polynomial in the variable x_{n+1} whose coefficients are polynomials in $\mathbb{F}[x_1, x_2, \ldots, x_n]$. As such, we can say that
$$\mathbb{F}[x_1, x_2, \ldots, x_n, x_{n+1}] = (\mathbb{F}[x_1, x_2, \ldots, x_n,])\, [x_{n+1}].$$
In the language of abstract algebra, although $\mathbb{F}[x_1, x_2, \ldots, x_n]$ is not a field, it is a ring (like $\mathbb{Z}$) and we can still define polynomials whose coefficients are from a ring.

Proof. We will show the method of turning a polynomial in $\mathbb{F}[x_1, x_2, \ldots, x_n, x_{n+1}]$ into a polynomial in $(\mathbb{F}[x_1, x_2, \ldots, x_n,])\, [x_{n+1}]$. This is best done by example. If we begin with
$$f(x, y) = 2x^3y^2 + 4x^2y + 5y + xy^2 \in \mathbb{R}[x, y]$$
and wish to write it as an element of $(\mathbb{R}[x])[y]$, we order the terms first by the exponents of y and then the exponents of x. This yields
$$f(x, y) = 2y^2x^3 + y^2x + 4yx^2 + 5y.$$
Finally, we consider x to be a constant and collect like terms in the variable y, which yields
$$f(x, y) = (2x^3 + x)y^2 + (4x^2 + 5)y.$$
This process generalizes. □

Theorem 11.49. The elementary symmetric polynomials in $\mathbb{F}[x_1, x_2, \ldots, x_n]$ are symmetric polynomials (thank goodness!).

Proof. Let the n variables be $x_1, x_2, \ldots, x_n$ and let x be an extra variable. Then we define
$$f \in \mathbb{F}[x_1, x_2, \ldots, x_n, x] = (\mathbb{F}[x_1, x_2, \ldots, x_n])\, [x]$$
$$f(x_1, x_2, \ldots, x_n, x) = (x - x_1)(x - x_2) \cdots (x - x_n).$$
Let $e_1, e_2, \ldots, e_n$ be the elementary symmetric polynomials in $x_1, x_2, \ldots, x_n$. By the same argument that we used when proving Vieta's formulas (Theorem 11.43), this expands as
$$x^n - e_1x^{n-1} + \cdots + (-1)^k e_k x^{n-k} + \cdots + (-1)^{n-1} e_{n-1} x + (-1)^n e_n.$$
Since swapping x_i with x_j in $(x - x_1)(x - x_2) \cdots (x - x_n)$ leaves the expression unchanged, it means that all of the coefficients of
$$x^n - e_1x^{n-1} + \cdots + (-1)^k e_k x^{n-k} + \cdots + (-1)^{n-1} e_{n-1} x + (-1)^n e_n$$
remain unchanged. By comparing coefficients, this simultaneously proves that all the e_k are symmetric polynomials. □

Theorem 11.50 (Fundamental theorem of symmetric polynomials). Every symmetric polynomial contained in $\mathbb{F}[x_1, x_2, \dots, x_n]$ can be uniquely written as a polynomial in terms of the elementary symmetric polynomials $e_1, e_2, \dots, e_k$.

Proof. There is an elementary proof of this result using lexicographical order (Definition 5.42) that is due to Gauss, but we will not delve into it as it is cumbersome and would not help with our purposes [1]. $\square$

Corollary 11.51. If we are given a polynomial

$$f(x) = a_n x^n + a_{n-1}x^{n-1} + \cdots + a_2x^2 + a_1x + a_0$$

in $\mathbb{C}[x]$ with unknown roots $r_1, r_2, \dots, r_n$. but known coefficients a_k, then we can compute any symmetric polynomial $g(r_1, r_2, \dots, r_n)$ in terms of the coefficients a_k.

Problem 11.52. Let $f(x, y, z) = 2x^3 + 3x^2 + 5x + 7$. Let its complex roots be r, s, t. Compute the value of the symmetric expression $g(r, s, t) = r^2 + s^2 + t^2$.

Theorem 11.53 (Girard-Newton sums). Let n be a positive integer and

$$p(x) = s_0x^n + s_1x^{n-1} + \cdots + s_{n-1}x + s_n$$

be a polynomial in $\mathbb{C}[x]$ with the complex roots $r_1, r_2, \dots, r_n$, where each root is included in this list as many times as its multiplicity in $p(x)$. For each positive integer k, let

$$p_k = r_1^k + r_2^k + \cdots + r_n^k,$$

with the convention

$$p_0 = \underbrace{1 + 1 + \cdots + 1}_{n \text{ copies of } 1} = n.$$

Then the following identities hold:

$$0 = \begin{cases} ks_k + \sum_{i=0}^{k-1} s_i p_{k-i} & \text{if } 1 \leq k < n \\ \sum_{i=0}^{n} s_i p_{k-i} & \text{if } 1 \leq n \leq k \end{cases}.$$

The coefficients in $p(x)$ have been strategically chosen to be indexed in reverse order so that these identities are easier to remember.

Proof. The identity for $k \geq n$ is easier to establish, and we will start with that. For each $j \in [n]$, r_j is a root of of $p(x)$, so

$$0 = p(r_j) = s_0r_j^n + s_1r_j^{n-1} + \cdots + s_{n-2}r_j^2 + s_{n-1}r_j + s_n.$$

Multiplying through by r_j^{k-n} (and using the temporary convention $0^0 = 1$, if needed), yields

$$0 = p(r_j) \cdot r_j^{k-n} = s_0r_j^k + s_1r_j^{k-1} + \cdots + s_{n-2}r_j^{k-n+2} + s_{n-1}r_j^{k-n+1} + s_nr_j^{k-n}.$$

By summing this equation over all $j \in [n]$, we get

$$0 = \sum_{j=1}^{n} p(r_j) \cdot r_j^{k-n} = \sum_{j=1}^{n}\sum_{i=0}^{n} s_ir_j^{k-i}.$$

By the discrete Fubini's principle, this is equivalent to

$$0 = \sum_{i=0}^{n}\sum_{j=1}^{n} s_ir_j^{k-i} = \sum_{i=0}^{n} s_i(r_1^{k-i} + r_2^{k-i} + \cdots + r_n^{k-i}) = \sum_{i=0}^{n} s_ip_{k-i},$$

which is the desired identity.
All proofs of the $k < n$ identity of which are aware are more complicated than the first case in one way or another. The cleanest one, in our opinion, is one that uses a tiny amount of calculus. We will show it here, as the derivative is an extremely useful tool to illustrate. The idea is to compute the derivative of $p(x)$ in two different ways and to equate corresponding coefficients. One way is standard, that is

$$p'(x) = ns_0x^{n-1} + (n-1)s_1x^{n-2} + \cdots + 2s_{n-2}x + s_{n-1} = \sum_{k=0}^{n-1}(n-k)s_kx^{n-k-1}.$$

The other way is to differentiate the factorization

$$p(x) = s_0(x-r_1)(x-r_2)\cdots(x-r_n).$$

By a generalization of the product rule from calculus via induction on $n \geq 2$,

$$p'(x) = \sum_{j=1}^{n} \frac{p(x)}{x-r_j}.$$

The summands do not actually have a discontinuity at $x = r_j$, but rather each $\dfrac{p(x)}{x-r_j}$ denotes the quotient upon $p(x)$ being perfectly divided by its factor $x - r_j$. Leaving the details to the reader (it is beneficial to write them out!), the quotient of $p(x)$ upon synthetic division by the linear factor $x - r_j$ is

$$\frac{p(x)}{x-r_j} = \sum_{k=0}^{n-1} x^{n-k-1}(s_0r_j^k + s_1r_j^{k-1} + \cdots + s_{k-1}r_j + s_k).$$

Summing this equation over all $j \in [n]$, we get

$$p'(x) = \sum_{j=1}^{n}\frac{p(x)}{x-r_j} = \sum_{j=1}^{n}\sum_{k=0}^{n-1} x^{n-k-1}(s_0r_j^k + s_1r_j^{k-1} + \cdots + s_{k-1}r_j + s_k).$$

By two applications of the discrete Fubini's principle, this is equivalent to

$$\begin{aligned}
p'(x) &= \sum_{j=1}^{n}\sum_{k=0}^{n-1} x^{n-k-1}\sum_{i=0}^{k} s_ir_j^{k-i} = \sum_{k=0}^{n-1}\sum_{j=1}^{n} x^{n-k-1}\sum_{i=0}^{k} s_ir_j^{k-i}\\
&= \sum_{k=0}^{n-1} x^{n-k-1}\sum_{j=1}^{n}\sum_{i=0}^{k} s_ir_j^{k-i} = \sum_{k=0}^{n-1} x^{n-k-1}\sum_{i=0}^{k}\sum_{j=1}^{n} s_ir_j^{k-i}\\
&= \sum_{k=0}^{n-1} x^{n-k-1}\sum_{i=0}^{k} s_i\sum_{j=1}^{n} r_j^{k-i} = \sum_{k=0}^{n-1} x^{n-k-1}\sum_{i=0}^{k} s_ip_{k-i}.
\end{aligned}$$

By comparing coefficients with

$$p'(x) = \sum_{k=0}^{n-1}(n-k)s_kx^{n-k-1},$$

we get for each $k \in [n-1]$ that

$$(n-k)s_k = \sum_{i=0}^{k} s_ip_{k-i} = \sum_{i=0}^{k-1} s_ip_{k-i} + s_kp_0 = \sum_{i=0}^{k-1} s_ip_{k-i} + s_kn.$$

Taking everything to one side and slightly simplifying yields

$$0 = ks_k + \sum_{i=0}^{k-1} s_ip_{k-i}.$$

This is what we wanted to prove. □

11.5 Multivariable Factoring

Now, we will see the analogue of the polynomial factor theorem (Theorem 11.27) for multivariable polynomials.

Theorem 11.54 (Multivariable factor theorem)**.** Let $n \geq 2$ be an integer. Suppose polynomials $f \in \mathbb{F}[x_1, x_2, \ldots, x_n]$ and $g \in \mathbb{F}[x_1, x_2, \ldots, x_{n-1}]$ exist such that

$$f(x_1, x_2, \ldots, x_{n-1}, g(x_1, x_2, \ldots, x_{n-1})) = 0.$$

Then $x_n - g$ is a factor of f in the sense that there exists a polynomial $h \in \mathbb{F}[x_1, x_2, \ldots, x_n]$ such that

$$f = (x_n - g)h.$$

The same result holds for any x_k instead of x_n.

Proof. The proof is most naturally done in the language of abstract algebra using commutative rings, so we will only briefly sketch it. There is a version of the Euclidean division theorem for polynomials (Theorem 11.23) that works on polynomials with coefficients in commutative rings instead of fields, with one caveat: the divisor has to be monic. We will only require the special case of a linear divisor, which is stated as follows:
If R is a commutative ring, then, for all $f \in R[x]$ and $g \in R$, there exists an $h \in R[x]$ and an $r \in R$ such that

$$f(x) = (x - g)h(x) + r.$$

The proof of it follows exactly the line of logic that we used in proving the Euclidean division theorem for polynomials (Theorem 11.23), so we do not repeat it here. We merely comment that it will work, despite the fact that the elements of R do not necessarily have multiplicative inverses, and it will work precisely because $x - g$ is monic. We encourage the reader to prove it, first for $f = 0$ and then by induction on $\deg f \geq 0$.
As a consequence, an analogue of the polynomial factor theorem is that $x - g$ divides f with no remainder (meaning $r = 0$) if and only if $f(g) = 0$.
Now we take R to be the polynomial ring $\mathbb{F}[x_1, x_2, \ldots, x_{n-1}]$, which is a commutative ring, and note that

$$\mathbb{F}[x_1, x_2, \ldots, x_{n-1}, x_n] = (\mathbb{F}[x_1, x_2, \ldots, x_{n-1}])[x_n] = R[x_n].$$

Suppose $f \in R[x_n]$ and $g \in R$ such that $f(g) = 0$. By our analogue of the polynomial factor theorem, there exists an $h \in R[x_n]$ such that $f = (x_n - g)h$, and we are done. □

Example 11.55. Factor $(x-y)^3 + (y-z)^3 + (z-x)^3$ into a product of a constant times linear factors.

Solution. Let

$$f(x, y, z) = (x-y)^3 + (y-z)^3 + (z-x)^3.$$

Note that each of the scenarios $x = y, y = z, z = x$ individually leads to $f(x, y, z) = 0$. This means each of the linear polynomials $x - y, y - z, z - x$ is a factor of f. So we can guess that there is a polynomial g such that

$$f(x, y, z) = (x-y)(y-z)(z-x)g(x, y, z).$$

Since $\deg f = 3$, we know that g must be a constant c. Now we can substitute constants into f, such as $(x, y, z) = (3, 2, 1)$ to get

$$c = \frac{(x-y)^3 + (y-z)^3 + (z-x)^3}{(x-y)(y-z)(z-x)} = \frac{1^3 + 1^3 + (-2)^3}{1 \cdot 1 \cdot (-2)} = 3.$$

Indeed, expanding allows us to check that

$$(x-y)^3 + (y-z)^3 + (z-x)^3 = 3(x-y)(y-z)(z-x).$$

Note that there was guesswork involved, since it was assumed that the quotient, upon dividing $f(x, y, z)$ by $x - y$, still has $y - z$ and $z - x$ as factors. □

Problem 11.56. Factor $(x+y+z)^3 - x^3 - y^3 - z^3$ into a product of linear factors times a constant.

We will end the section by discussing some common multivariable factorizations and listing many others. In general, we like expansions of factored expressions where the simplified expression, after like terms have been collected, has very few terms.

Theorem 11.57 (Difference or sum of powers). Recall the difference of squares factorization

$$x^2 - y^2 = (x-y)(x+y).$$

This generalizes to all integers $n \geq 1$, as follows:

1. $x^{n+1} - y^{n+1} = (x-y)(x^n + x^{n-1}y + x^{n-2}y^2 + \cdots + x^2y^{n-2} + xy^{n-1} + y^n)$
2. $x^{2n+1} + y^{2n+1} = (x+y)(x^{2n} - x^{2n-1}y + x^{2n-2}y^2 - \cdots + x^2y^{2n-2} - xy^{2n-1} + y^{2n})$

Proof. We will prove the difference of powers factorization using geometric series, and then we will prove the sum of odd powers factorization using the difference of powers.

1. Although a direct expansion with telescoping would work, let's try something more interesting. The formula for a geometric series (Theorem 9.14) says that, for $z \neq 1$.

$$1 + z + z^2 + \cdots + z^n = \frac{z^{n+1} - 1}{z - 1}.$$

   This is equivalent to

$$z^{n+1} - 1 = (z-1)(z^n + z^{n-1} + \cdots + z + 1).$$

   We can check separately that this new equation holds for $z = 1$ as well. Letting $z = \dfrac{x}{y}$ and multiplying both sides by y^{n+1} yields the desired factorization. Since we can separately check that $y = 0$ works with ease, the difference of powers factorization holds for all complex numbers x and y.

2. Since $2n+1$ is odd,

$$x^{2n+1} + y^{2n+1} = x^{2n+1} - (-y)^{2n+1}.$$

   Then we simply apply the difference of powers formula from the first part to the right side.

□

Problem 11.58. There are multiple ways of applying the sum or difference of powers factorizations to composite exponents, as follows:

$$x^{mn} \pm y^{mn} = (x^m)^n \pm (y^m)^n = (x^n)^m \pm (y^n)^m.$$

Use this idea to factor $x^6 - y^6$ into irreducible factors with real coefficients.

Problem 11.59. Find a quadratic with real coefficients that has a non-trivial third root of unity as a root. By non-trivial, we mean that it is not equal to 1.

Problem 11.60. The following is a very common lemma. Suppose P is a polynomial with integer coefficients and a, b are integers such that $a \neq b$. Show that the integer $a - b$ divides the integer $P(a) - P(b)$. Deduce that if b is a root of P, then $a - b$ divides $P(a)$.

Example 11.61 (Sophie Germain identity). Factor $x^4 + 4y^4$ as the product of two polynomials, each of which is a sum of two squares.

Solution. We can rewrite the expression as follows using the difference of squares:

$$\begin{aligned} x^4 + 4y^4 &= (x^2)^2 + (2y^2)^2 + 4x^2y^2 - 4x^2y^2 \\ &= (x^2 + 2y^2)^2 - (2xy)^2 \\ &= (x^2 - 2xy + 2y^2)(x^2 + 2xy + 2y^2) \\ &= ((x-y)^2 + y^2)((x+y)^2 + y^2). \end{aligned}$$

□

Theorem 11.62. As a reference, below is a master list of additional well-known manipulations in two or three variables. The reader is encouraged to verify them independently.

$$
\begin{aligned}
x^4 - y^4 &= (x-y)(x+y)(x^2+y^2) \\
x^4 + 4y^4 &= ((x-y)^2+y^2)((x+y)^2+y^2) \\
x^4 + x^2y^2 + y^4 &= (x^2+xy+y^2)(x^2-xy+y^2) \\
x^4 + y^4 + (x \pm y)^4 &= 2(x^2 \pm xy + y^2)^2 \\
(x+y)^2 - x^2 - y^2 &= 2xy \\
(x+y)^3 - x^3 - y^3 &= 3xy(x+y) \\
(x+y)^5 - x^5 - y^5 &= 5xy(x+y)(x^2+xy+y^2) \\
(x+y)^7 - x^7 - y^7 &= 7xy(x+y)(x^2+xy+y^2)^2 \\
(x+y+z)^2 - x^2 - y^2 - z^2 &= 2(xy+yz+zx) \\
(x+y+z)^3 - x^3 - y^3 - z^3 &= 3(x+y)(y+z)(z+x) \\
(x+y+z)^5 - x^5 - y^5 - z^5 &= 5(x+y)(y+z)(z+x)(x^2+y^2+z^2+xy+yz+zx) \\
\frac{(x+y+z)^7 - x^7 - y^7 - z^7}{(x+y)(y+z)(z+x)} &= 7((x^2+y^2+z^2+xy+yz+zx)^2 + xyz(x+y+z)) \\
(xy^2+yz^2+zx^2)(x^2y+y^2z+z^2x) - x^2y^2z^2 &= (x^2+yz)(y^2+zx)(z^2+xy) \\
(xy^2+yz^2+zx^2-xyz)^2 + (x^2y+y^2z+z^2x-xyz)^2 &= (x^2+y^2)(y^2+z^2)(z^2+x^2) \\
(x+y+z)(xy+yz+zx) - xyz &= (x+y)(y+z)(z+x) \\
x^3+y^3+z^3-3xyz &= (x+y+z)(x^2+y^2+z^2-xy-yz-zx) \\
(x+y+z)^3 - 24xyz &= (x+y+z)^3 + (x-y+z)^3 + (x+y-z)^3 \\
x^3+y^3+z^3+(x+y)^3+(y+z)^3+(z+x)^3 &= 3(x+y+z)(x^2+y^2+z^2) \\
(x-y)^3+(y-z)^3+(z-x)^3 &= 3(x-y)(y-z)(z-x) \\
(xy^2+yz^2+zx^2) - (x^2y+y^2z+z^2x) &= (x-y)(y-z)(z-x) \\
(xy^3+yz^3+zx^3) - (x^3y+y^3z+z^3x) &= (x+y+z)(x-y)(y-z)(z-x) \\
(x^2y^3+y^2z^3+z^2x^3) - (x^3y^2+y^3z^2+z^3x^2) &= (xy+yz+zx)(x-y)(y-z)(z-x)
\end{aligned}
$$

Problem 11.63. Let a, b, c be complex numbers such that $a+b+c=0$. For each positive integer n, let

$$f(n) = \frac{a^n+b^n+c^n}{n}.$$

Prove that the following two identities hold: $f(2) \cdot f(3) = f(5)$ and $f(2) \cdot f(5) = f(7)$.

Chapter 12

Multivariable Inequalities

"A mathematician who is not also something of a poet will never be a complete mathematician."

– *Karl Weierstrass*

"Good, [the student who dropped out to study poetry] did not have enough imagination to become a mathematician."

– *David Hilbert*

"Pure mathematics is, in its way, the poetry of logical ideas."

– *Albert Einstein*

Multivariable inequalities is an area of elementary mathematics that saw tremendous growth the first two decades of the twenty-first century. Entire books could be (and have been) written about them, especially by certain Vietnamese and Romanian practitioners. The advent of advanced techniques that allow for proofs of large classes of inequalities has resulted in problems about inequalities appearing less frequently on competitions and olympiads, and it has also caused the production of unnatural and less beautiful inequalities in the problems literature. We will focus only on classical results with which we feel the reader should be familiar.

12.1 AM-GM and Cauchy-Schwarz

An optimization result generally has two parts: a bound and statement about where equality is achieved in the bound, if at all. If we are lucky, it is possible to make a biconditional statement (recall that this means "if and only if") about when equality occurs. In some cases, there is no known concise presentation of the equality cases, so only sufficient or necessary conditions might be stated. If the inequality is ugly enough, there might be no mention of equality criteria at all.

Problem 12.1. Let a, b, c be real numbers. Then prove that the following classic inequalities hold and find where equality occurs in each case:

$$a^2 + b^2 \geq 2ab$$
$$a^2 + b^2 + c^2 \geq ab + bc + ca.$$

Definition 12.2. For any positive integer n and non-negative real numbers $a_1, a_2, \ldots, a_n$, their **arithmetic mean** (AM for short) is

$$\frac{a_1 + a_2 + \cdots + a_n}{n}$$

and their **geometric mean** (GM for short) is

$$\sqrt[n]{a_1 a_2 \cdots a_n}.$$

Theorem 12.3 (AM-GM inequality). Let n be a positive integer and $a_1, a_2, \ldots, a_n$ be non-negative real numbers. Then

$$\frac{a_1 + a_2 + \cdots + a_n}{n} \geq \sqrt[n]{a_1 a_2 \cdots a_n},$$

with equality holding if and only if all of the a_i are equal.

Proof. We will use a special form of induction called Cauchy induction or forward-backward induction. Let $S(n)$ denote the assertion that

$$\frac{a_1 + a_2 + \cdots + a_n}{n} \geq \sqrt[n]{a_1 a_2 \cdots a_n}$$

for all non-negative a_i, with equality holding if and only if all the a_i are equal. It will suffice to prove the following in sequence:

1. $S(2)$, which establishes the base case ($S(1)$ is trivial and it can be proven alone)

2. $S(n) \implies S(2n)$, which establishes $S(2^k)$ for all positive integers k when repeatedly applied, starting with $S(2)$

3. $S(n) \implies S(n-1)$, which establishes $S(n)$ for all remaining n since we can work backwards from each $S(2^k)$

Now we will prove that these three hold:

1. We already know that $S(2)$ is true, due to the trivial inequality (Problem 12.1).

2. Suppose $S(n)$ holds. Then we use $S(2)$ and $S(n)$ to get

$$\begin{aligned}\frac{a_1 + a_2 + \cdots + a_{2n}}{2n} &= \frac{1}{2}\left(\frac{a_1 + a_2 + \cdots + a_n}{n} + \frac{a_{n+1} + a_{n+2} + \cdots + a_{2n}}{n}\right) \\ &\geq \frac{1}{2}\left(\sqrt[n]{a_1 a_2 \cdots a_n} + \sqrt[n]{a_{n+1} a_{n+1} \cdots a_{2n}}\right) \\ &\geq \sqrt[2n]{a_1 a_2 \cdots a_{2n}}.\end{aligned}$$

   We can derive that equality holds if and only if all the a_i are equal using the equality conditions for $S(2)$ and $S(n)$.

3. Now suppose $S(n)$ holds. Miraculously, we find that we can rewrite the arithmetic mean in $S(n-1)$ as follows, and then we can apply $S(n)$ to get

$$\begin{aligned}\frac{a_1 + a_2 + \cdots + a_{n-1}}{n-1} &= \frac{1}{n}\left(a_1 + a_2 + \cdots + a_{n-1} + \frac{a_1 + a_2 + \cdots + a_{n-1}}{n-1}\right) \\ &\geq \sqrt[n]{a_1 a_2 \cdots a_{n-1} \cdot \frac{a_1 + a_2 + \cdots + a_{n-1}}{n-1}}.\end{aligned}$$

   Upon taking the nth power of each side, cancelling a copy of $\frac{a_1 + a_2 + \cdots + a_{n-1}}{n-1}$ from both sides, and finally taking the $(n-1)^{\text{th}}$ root of each side, this becomes the inequality for $S(n-1)$. The equality condition for $S(n-1)$ follows from the equality condition for $S(n)$.

□

Problem 12.4. In the third step of the proof of Theorem 12.3, we used the substitution

$$a_n = \frac{a_1 + a_2 + \cdots + a_{n-1}}{n-1}$$

to show that $S(n) \implies S(n-1)$. Use the substitution

$$a_n = \sqrt[n-1]{a_1 a_2 \cdots a_{n-1}}$$

to provide an alternative proof of this implication.

Example 12.5. Prove that, for all real numbers $r \geq -1$ and positive integers $m > n$, it holds that

$$\left(1 + \frac{r}{m}\right)^m \geq \left(1 + \frac{r}{n}\right)^n,$$

with equality holding if and only if $r = 0$. This goes hand in hand with Bernoulli's inequality (Theorem 5.36) because, where Bernoulli shows that compound interest is better than simple interest for the receiver, this shows that more compounding at the same rate means the deal keeps getting better for the receiver. What do you think happens as $n \to \infty$?

Solution. Let $r \geq -1$ be a real number and n be a positive integer. We will show that

$$\left(1+\frac{r}{n+1}\right)^{n+1} \geq \left(1+\frac{r}{n}\right)^{n},$$

with equality holding if and only if $r = 0$. It suffices to prove this instead because we can repeatedly apply this result to compare the expressions for m and n instead, where $m > n$.
First, note that

$$r \geq -1 \implies n \geq 1 \geq -r \implies \frac{r}{n} \geq -1 \implies 1+\frac{r}{n} \geq 0.$$

So we can apply the AM-GM inequality to one copy of 1 and n copies of $1+\frac{r}{n}$ to get

$$\frac{1+n\cdot\left(1+\frac{r}{n}\right)}{n+1} \geq \sqrt[n+1]{1\cdot\left(1+\frac{r}{n}\right)^{n}}.$$

This implies the comparison between the expressions for $n+1$ and n that we wanted to see. By the equality criterion for the AM-GM inequality, equality holds if and only if

$$1 = 1+\frac{r}{n} \iff r = 0,$$

as hypothesized.
Interestingly, the benefits of more compounding taper off. It is known that

$$\lim_{n\to\infty}\left(1+\frac{r}{n}\right)^{n} = e^{r},$$

where e is Euler's constant and r is any real number. □

Problem 12.6 (GM-HM inequality). Let n be a positive integer and $a_1, a_2, \ldots, a_n$ be positive real numbers. Show that

$$\sqrt[n]{a_1 a_2 \cdots a_n} \geq \frac{n}{\frac{1}{a_1}+\frac{1}{a_2}+\cdots+\frac{1}{a_n}},$$

with equality holding if and only if all of the a_i are equal. The quantity on the right side is called the **harmonic mean** (HM for short) of the numbers, hence the name GM-HM inequality.

Problem 12.7. Solve the following two optimization problems.

1. If p is a positive real number, then determine the dimensions of a rectangle with maximal area among all rectangles with perimeter p. The dimensions should be written in terms of p.

2. If S is a positive real number, then determine the dimensions of a box with maximal volume among all boxes with surface area S. The dimensions should be written in terms of S.

Problem 12.8. The AM-GM inequality states that

$$\frac{x^3+y^3+z^3}{3} \geq xyz$$

for all non-negative real numbers x, y, z. Determine all triples (x, y, z) of *real* numbers (not just non-negative ones) for which this inequality holds and state a biconditional equality condition.

Theorem 12.9 (Cauchy-Schwarz inequality). Suppose n is a positive integer. For all real numbers

$$a_1, a_2, \ldots, a_n \text{ and } b_1, b_2, \ldots, b_n,$$

it holds that

$$(a_1^2+a_2^2+\cdots+a_n^2)(b_1^2+b_2^2+\cdots+b_n^2) \geq (a_1b_1+a_2b_2+\cdots+a_nb_n)^2.$$

Equality holds if and only if either $a_i = 0$ for all $i \in [n]$ or there exists a real constant r such that $a_i r = b_i$ for all $i \in [n]$.

Proof. If we rearrange the inequality and multiply through by 4, we get the equivalent inequality

$$0 \geq (2a_1b_1 + 2a_2b_2 + \cdots + 2a_nb_n)^2 - 4(a_1^2 + a_2^2 + \cdots + a_n^2)(b_1^2 + b_2^2 + \cdots + b_n^2),$$

where the right side looks like a discriminant. This inspires to construct a quadratic function $f : \mathbb{R} \to \mathbb{R}$ defined by

$$f(x) = (a_1^2 + a_2^2 + \cdots + a_n^2)x^2 - (2a_1b_1 + 2a_2b_2 + \cdots + 2a_nb_n)x + (b_1^2 + b_2^2 + \cdots + b_n^2),$$

which has the desired discriminant. At this point, we should note that this is not a quadratic function if all the a_i are equal to 0 because then the quadratic term would disappear. However, it is easy to see that the Cauchy-Schwarz inequality holds with both sides being equal in this case. So, now we can assume that at least one of the a_i is non-zero.
By rearranging the expression for f so that terms with the same indices are grouped together, we see that it can be rewritten as

$$\begin{aligned} f(x) &= (a_1^2x^2 - 2a_1b_1x + b_1^2) + (a_2^2x^2 - 2a_2b_2x + b_2^2) + \cdots + (a_n^2x^2 - 2a_nb_nx + b_n^2) \\ &= (a_1x - b_1)^2 + (a_2x - b_2)^2 + \cdots + (a_nx - b_n)^2 \geq 0. \end{aligned}$$

The fact that f is always non-negative means that there is either one real solution or no real solutions. This is equivalent to the discriminant being non-positive, which proves the desired inequality.
So, equality holds if and only if all of the a_i are equal to 0, or the discriminant is 0. In the latter case, this is equivalent to there existing a real root r of f (two distinct real roots is not possible for f, so the existence of a real root is equivalent to the existence of exactly one real root). We found that

$$f(x) = (a_1x - b_1)^2 + (a_2x - b_2)^2 + \cdots + (a_nx - b_n)^2,$$

so there exists a real root r if and only if

$$(a_1r - b_1)^2 + (a_2r - b_2)^2 + \cdots + (a_nr - b_n)^2 = 0.$$

By the equality condition of the trivial inequality, this is true if and only if there exists a real number r such that $a_ir = b_i$ for all i. □

Corollary 12.10. Let n be a positive integer. The following are special cases of the Cauchy-Schwarz inequality that are useful in practice.

1. If $a_1, a_2, \ldots, a_n$ are positive real numbers, then

$$(a_1 + a_2 + \cdots + a_n)\left(\frac{1}{a_1} + \frac{1}{a_2} + \cdots + \frac{1}{a_n}\right) \geq n^2,$$

with equality holding if and only if all of the a_i are equal.

2. If $b_1, b_2, \ldots, b_n$ are positive real numbers and $a_1, a_2, \ldots, a_n$ are non-negative real numbers, then

$$(a_1b_1 + a_2b_2 + \cdots + a_nb_n)\left(\frac{a_1}{b_1} + \frac{a_2}{b_2} + \cdots + \frac{a_n}{b_n}\right) \geq (a_1 + a_2 + \cdots + a_n)^2.$$

If the a_i are all positive, then equality holds if and only if all the b_i are equal.

3. If $a_1, a_2, \ldots, a_n$ are positive real numbers and $b_1, b_2, \ldots, b_n$ are real numbers, then

$$(a_1 + a_2 + \cdots + a_n)\left(\frac{b_1^2}{a_1} + \frac{b_2^2}{a_2} + \cdots + \frac{b_n^2}{a_n}\right) \geq (b_1 + b_2 + \cdots + b_n)^2,$$

with equality holding if and only if there exists a real number r such that $a_ir = b_i$ for all i. This inequality has several names, one of which is **Engel's form**.

Proof. These are all applications of the Cauchy-Schwarz inequality.

1. Let $a_1, a_2, \ldots, a_n$ be positive real numbers. Applying the Cauchy-Schwarz inequality to the tuples

$$(\sqrt{a_1}, \sqrt{a_2}, \ldots, \sqrt{a_n}) \text{ and } \left(\frac{1}{\sqrt{a_1}}, \frac{1}{\sqrt{a_2}}, \ldots, \frac{1}{\sqrt{a_n}}\right)$$

yields the desired inequality. Since the a_i are positive, we do not have to worry about the equality case where all of the a_i are zero. Equality holds if and only if there exists a real number r such that $\sqrt{a_i} \cdot r = \frac{1}{\sqrt{a_i}}$ for all i, which is equivalent to all of the a_i being equal to each other and to $\frac{1}{r}$.

2. Let $b_1, b_2, \ldots, b_n$ be positive real numbers and $a_1, a_2, \ldots, a_n$ be non-negative real numbers. Applying Cauchy-Schwarz to the tuples

$$\left(\sqrt{a_1 b_1}, \sqrt{a_2 b_2}, \ldots, \sqrt{a_n b_n}\right) \text{ and } \left(\sqrt{\frac{a_1}{b_1}}, \sqrt{\frac{a_2}{b_2}}, \ldots, \sqrt{\frac{a_n}{b_n}}\right)$$

yields the desired identity. If all of the a_i are positive, then the $\sqrt{a_i b_i}$ are all positive and so equality holds if and only if there exists a real number r such that $\sqrt{a_i b_i} \cdot r = \sqrt{\frac{a_i}{b_i}}$ for all i. This equality condition holds if and only if all of the b_i are equal to each other and to $\frac{1}{r}$.

3. Let $a_1, a_2, \ldots, a_n$ be positive real numbers and $b_1, b_2, \ldots, b_n$ be real numbers. Applying Cauchy-Schwarz to the tuples

$$(\sqrt{a_1}, \sqrt{a_2}, \ldots, \sqrt{a_n}) \text{ and } \left(\frac{b_1}{\sqrt{a_1}}, \frac{b_2}{\sqrt{a_2}}, \ldots, \frac{b_n}{\sqrt{a_n}}\right)$$

yields the desired inequality. Since the a_i are all positive, equality holds if and only if there exists a real number r such that

$$\sqrt{a_i} \cdot r = \frac{b_i}{\sqrt{a_i}}$$

for all i. This equality condition is true if and only if there exists a real number r such that $a_i r = b_i$ for all i.

□

Problem 12.11. If $\sigma : [n] \to [n]$ is a bijection and $a_1, a_2, \ldots, a_n$ are real numbers, then show that

$$a_1^2 + a_2^2 + \cdots + a_n^2 \geq |a_1 a_{\sigma(1)} + a_2 a_{\sigma(2)} + \cdots + a_n a_{\sigma(n)}|.$$

Problem 12.12 (RMS-AM inequality). Let n be a positive integer and $a_1, a_2, \ldots, a_n$ be real numbers. Show that

$$n \cdot (a_1^2 + a_2^2 + \cdots + a_n^2) \geq (a_1 + a_2 + \cdots + a_n)^2,$$

and as a result,

$$\sqrt{\frac{a_1^2 + a_2^2 + \cdots + a_n^2}{n}} \geq \frac{a_1 + a_2 + \cdots + a_n}{n},$$

with equality holding if and only if all of the a_i are non-negative and equal. The expression on the left side is called the **root mean square** (RMS for short) of the numbers, hence the name RMS-AM inequality.

Problem 12.13 (Nesbitt's inequality). Prove that, for all positive real numbers a, b, c,

$$\frac{a}{b+c} + \frac{b}{c+a} + \frac{c}{a+b} \geq \frac{3}{2},$$

and show that equality holds if and only if $a = b = c$.

12.2 Schur

Theorem 12.14 (Schur's inequality). For all non-negative real numbers x, y, z and all positive real numbers t, it holds that

$$x^t(x-y)(x-z)+y^t(y-z)(y-x)+z^t(z-x)(z-y) \geq 0.$$

Equality holds if and only if $x=y=z$, or two of x, y, z are equal to each other with the third equal to 0. If t is a non-negative even integer (with the convention that $0^0=1$), then the inequality holds for all real x, y, z and the equality condition is the same.

Proof. Let t be a positive real number. Define the three-variable polynomial

$$P(x, y, z)=x^t(x-y)(x-z)+y^t(y-z)(y-x)+z^t(z-x)(z-y).$$

Since P is symmetric in the variables x, y, z, proving that the inequality holds when $x \geq y \geq z$ will prove it for all other orderings, and so all real triples x, y, z. So, we may assume without loss of generality that $x \geq y \geq z$. We may rewrite

$$P(x, y, z)=(x-y)(x^t(x-z)-y^t(y-z))+z^t(x-z)(y-z).$$

Using the ordering $x \geq y \geq z$, the following four inequalities hold for all real x, y, z:

$$\begin{aligned} x-y &\geq 0, \\ x-z &\geq y-z, \\ x-z &\geq 0, \\ y-z &\geq 0. \end{aligned}$$

If x, y, z are non-negative, then $x^t \geq y^t$ as well for all positive real t, thus proving that the above expression for $P(x, y, z)$ must be non-negative.
Alternatively, suppose x, y, z are any real numbers and t is an even positive integer (the $t=0$ case is easy to verify separately). We will show that $P(x, y, z)$ is non-negative by doing casework on the sign of y. The four inequalities above still hold and will be useful.

- If $y=0$, then we can show that

  $$P(x, y, z)=(x-z)(x^{t+1}-z^{t+1}),$$

  which must be non-negative because $x \geq y \geq z$ implies that $x \geq 0$ and $0 \geq z$, and $t+1$ is an odd positive integer.

- If $y>0$, then $x>0$ too. Then

  $$P(x, y, z)=(x-y)(x^t(x-z)-y^t(y-z))+z^t(x-z)(y-z)$$

  is non-negative because $x^t \geq y^t$, and $z^t \geq 0$ as well since t is an even positive integer.

- If $y<0$, then $z<0$ too. Then

  $$P(x, y, z)=x^t(x-y)(x-z)+(y-z)(z^t(x-z)-y^t(x-y))$$

  is non-negative because $z^t \geq y^t$, and $x^t \geq 0$ as well since t is an even positive integer.

Thus, Schur's inequality holds for all real x, y, z if t is an even positive integer.
Now we have to prove the biconditional equality criterion, which can be stated in logical notation as

$$(x=y \wedge y=z \wedge z=x) \vee(x=y \wedge z=0) \vee(y=z \wedge x=0) \vee(z=x \wedge y=0).$$

It is easy to show that, if any one of the four components of this logical disjunction holds, then $P(x, y, z)=0$. Now suppose the negation of the equality condition holds:

$$(x \neq y \vee y \neq z \vee z \neq x) \wedge(x \neq y \vee z \neq 0) \wedge(y \neq z \vee x \neq 0) \wedge(z \neq x \vee y \neq 0).$$

First we break this up into two disjoint cases: x, y, z are all distinct or at least two of x, y, z are equal. We can further break the latter case into three cases: $x = y$ or $y = z$ or $z = x$, which are three disjoint cases because any two of the equalities holding would cause $x = y = z$, contradicting the condition $(x \neq y) \vee (y \neq z) \vee (z \neq x)$ in the negation. In the $x = y$ case, the negated condition becomes

$$(y \neq z \vee z \neq x) \wedge (z \neq 0) \wedge (y \neq z \vee x \neq 0) \wedge (z \neq x \vee y \neq 0).$$

It must be true that both $y \neq z$ and $z \neq x$, otherwise $x = y = z$ is true, which we have already stated to be contradictory. So the $x = y$ case of the negation is equivalent to

$$(x = y) \wedge (x \neq z) \wedge (y \neq z) \wedge (z \neq 0).$$

Similarly, the $y = z$ case is equivalent to

$$(y = z) \wedge (y \neq x) \wedge (z \neq x) \wedge (x \neq 0)$$

and the $z = x$ case is equivalent to

$$(z = x) \wedge (z \neq y) \wedge (x \neq y) \wedge (y \neq 0).$$

We will show that each of these four cases separately lead to $P(x, y, z) > 0$. The reader should check that the arguments below simultaneously hold in the case where x, y, and z are non-negative real numbers with t a positive real number, and the case where x, y, z are any real numbers with t an even positive integer.

- In the case that x, y, z are all distinct, we will give an argument for the strict ordering $x > y > z$ which may be adapted to any other strict ordering, since P is symmetric in x, y, z. If $x > y > z$, then

$$\begin{aligned} P(x, y, z) &= (x - y)(x^t(x - z) - y^t(y - z)) + z^t(x - z)(y - z) \\ &\geq (x - y)(x^t(x - z) - y^t(y - z)) \\ &> (x - y)(x^t(y - z) - y^t(y - z)) \\ &= (x - y)(y - z)(x^t - y^t) \\ &> 0. \end{aligned}$$

- In the $x = y$ case,

$$P(x, y, z) = z^t(z - x)(z - y) = z^t(z - x)^2,$$

which must be positive because $z \neq 0$ and $x \neq z$. The $y = z$ and $z = x$ cases follow symmetrically.

Thus, equality cannot hold in any of the four cases, so the stated equality condition is biconditional. □

Corollary 12.15. The $t = 0, 1, 2, 3$ cases of Schur's inequality can appear in various disguised ways using special factorizations of multivariable polynomials. For $t = 0$, we get the classic

$$x^2 + y^2 + z^2 \geq xy + yz + zx.$$

For $t = 1$, a factored version is

$$xyz \geq (-x + y + z)(x - y + z)(x + y - z)$$

and an equivalent form that is stronger than Nesbitt's inequality (Problem 12.13) is

$$\frac{x}{y+z} + \frac{y}{z+x} + \frac{z}{x+y} + \frac{4xyz}{(x+y)(y+z)(z+x)} \geq 2.$$

A factored form of $t = 2$ is

$$(x + y + z)(x^3 + y^3 + z^3 + 3xyz) \geq 2(x^2 + y^2 + z^2)(xy + yz + zx).$$

For $t = 3$, we find that

$$x^2 + y^2 + z^2 + \frac{6xyz}{x+y+z} + \frac{(x+y+z)xyz}{x^2+y^2+z^2} \geq 2(xy + yz + zx).$$

Problem 12.16 (Euler's inequality). Prove that, for every triangle, if R is its circumradius and r is its inradius, then $R \geq 2r$, and show that equality holds if and only if the triangle is equilateral. It will be helpful to note that

$$R = \frac{abc}{\sqrt{(a+b+c)(-a+b+c)(a-b+c)(a+b-c)}},$$
$$r = \frac{1}{2} \cdot \sqrt{\frac{(-a+b+c)(a-b+c)(a+b-c)}{a+b+c}},$$

where a, b, c are the side lengths of the triangle. A different short proof that uses excircles was published by the author of this book [9].

Example 12.17 (Gerretsen's inequalities). Prove that, for every triangle, if R is its circumradius, r is its inradius and s is its semiperimeter, then

$$16Rr - 5r^2 \leq s^2 \leq 4R^2 + 4Rr + 3r^2,$$

with equality holding in either inequality if and only if the triangle is equilateral. Note that the semiperimeter is defined as half of the perimeter.

Solution. There is a technique called Ravi substitution from Volume 4, which defines

$$x = s - a,$$
$$y = s - b,$$
$$z = s - c.$$

Thanks to the triangle inequality, x, y, z are all positive real numbers. We can solve for a, b, c to get

$$a = y + z,$$
$$b = z + x,$$
$$c = x + y.$$

Relevant formulas for R and r were stated in Problem 12.16, in terms of the side lengths a, b, c of the triangle. Now we can rewrite the formulas for R, r, s in terms of x, y, z instead of a, b, c:

$$R = \frac{(x+y)(y+z)(z+x)}{4\sqrt{(x+y+z)xyz}},$$
$$r = \sqrt{\frac{xyz}{x+y+z}},$$
$$s = x + y + z.$$

Substituting these into $16Rr - 5r^2 \leq s^2$ and clearing the denominators yields

$$4(x+y)(y+z)(z+x) - 5xyz \leq (x+y+z)^3,$$

which turns out to be another way of writing Schur inequality for $t = 1$, that is

$$x(x-y)(x-z) + y(y-z)(y-x) + z(z-x)(z-y) \geq 0.$$

Since none of x, y, z can be 0, equality holds if and only if $x = y = z$, which is true if and only if $a = b = c$. The other side of this double inequality is messier. Without showing the details, we can substitute the formulas for R, r, s from Problem 12.16 into $s^2 \leq 4R^2 + 4Rr + 3r^2$, clear the denominators, expand every term, and rearrange them to produce the equivalent inequality

$$x^4(y-z)^2 + y^4(z-x)^2 + z^4(x-y)^2 + 2(u(u-v)(u-w) + v(v-w)(v-u) + w(w-u)(w-v)) \geq 0,$$

where

$$u = xy, v = yz, w = zx.$$

The first three terms are non-negative by the trivial inequality and the last three terms together are non-negative by Schur's inequality for $t = 1$. Again, since none of x, y, z can be 0, equality holds if and only if $x = y = z$, which is true if and only if the triangle is equilateral. The reader should work out the expansion manually as performing such computations is not uncommon, and so it is a good skill to practice. □

Problem 12.16 and Example 12.17 not just curiosities. Euler's inequality and Gerretsen's inequalities are the backbone of many proofs that use the powerful Rrs technique of proving geometric inequalities about triangles. This is a method that expresses all terms in a geometric inequality purely in terms of R, r, s (as opposed to other quantities, like the side lengths a, b, c), and then utilizes Euler or Gerretsen, which are usually strong enough. In sharper cases, there is a monster available, called Blundon's inequalities.

12.3 Rearrangement and Chebyshev

Theorem 12.18 (Rearrangement inequality)**.** Suppose n is a positive integer and suppose

$$(x_1, x_2, \dots, x_n) \text{ and } (y_1, y_2, \dots, y_n)$$

are non-decreasing n-tuples of real numbers. By non-decreasing, we mean that

$$x_1 \le x_2 \le \cdots \le x_n,$$
$$y_1 \le y_2 \le \cdots \le y_n.$$

Then, for every bijection $\sigma : [n] \to [n]$, it holds that

$$\sum_{i=1}^{n} x_i y_{n-i+1} \le \sum_{i=1}^{n} x_i y_{\sigma(i)} \le \sum_{i=1}^{n} x_i y_i.$$

We have not seen an equality condition for the rearrangement inequality that holds biconditionally, but the following sufficient criteria hold. If $(x_1, x_2, \dots, x_n)$ and $(y_1, y_2, \dots, y_n)$ are strictly increasing instead of merely non-decreasing, meaning

$$x_1 < x_2 < \cdots < x_n,$$
$$y_1 < y_2 < \cdots < y_n,$$

then equality holds in the right inequality if and only if $\sigma(i) = i$ for all i, and equality holds in the left inequality if and only if $\sigma(i) = n - i + 1$ for all i. Regardless of whether the strict inequalities hold, another sufficient equality criterion is that at least one of the n-tuples $(x_1, x_2, \dots, x_n)$ and $(y_1, y_2, \dots, y_n)$ consists of all equal entries.

Proof. We will first prove the following assertion by induction on positive integers n: For all n-tuples of real numbers

$$(x_1, x_2, \dots, x_n) \text{ and } (y_1, y_2, \dots, y_n)$$

and all bijections $\sigma : [n] \to [n]$, it holds that

$$\sum_{i=1}^{n} x_i y_{\sigma(i)} \le \sum_{i=1}^{n} x_i y_i,$$

with equality holding in the case of

$$(x_1, x_2, \dots, x_n) \text{ and } (y_1, y_2, \dots, y_n)$$

being strictly increasing if and only if σ is the identity map on $[n]$.

In the base case $n = 1$, the only bijection $\sigma : [1] \to [1]$ is the identity map $\sigma(1) = 1$. The result clearly holds in this case. Now suppose the result holds for some positive integer n. Let

$$(x_1, x_2, \ldots, x_n, x_{n+1}) \text{ and } (y_1, y_2, \ldots, y_n, y_{n+1})$$

be non-decreasing $(n+1)$-tuples of real numbers, and let $\sigma : [n+1] \to [n+1]$ be a bijection. We split the argument into two cases: $\sigma(n+1) = n+1$ and $\sigma(n+1) \neq n+1$.
If $\sigma(n+1) = n+1$, then we can cancel $x_{n+1}y_{\sigma(n+1)} = x_{n+1}y_{n+1}$ from both sides of the desired inequality

$$\sum_{i=1}^{n+1} x_i y_{\sigma(i)} \leq \sum_{i=1}^{n+1} x_i y_i$$

so that it is equivalent to prove that

$$\sum_{i=1}^{n} x_i y_{\sigma(i)} \leq \sum_{i=1}^{n} x_i y_i.$$

This follows from invoking the induction hypothesis because the restriction $\sigma|_{[n]}$ is a bijection from $[n]$ to $[n]$ in this case, and the non-decreasing property is preserved. For the sufficient equality condition, suppose

$$(x_1, x_2, \ldots, x_n, x_{n+1}) \text{ and } (y_1, y_2, \ldots, y_n, y_{n+1})$$

are strictly increasing. By the induction hypothesis, equality holds in

$$\sum_{i=1}^{n} x_i y_{\sigma(i)} \leq \sum_{i=1}^{n} x_i y_i$$

if and only if $\sigma|_{[n]}$ is the identity map on $[n]$. Since we have assumed that $\sigma(n+1) = n+1$, equality holds in

$$\sum_{i=1}^{n+1} x_i y_{\sigma(i)} \leq \sum_{i=1}^{n+1} x_i y_i$$

if and only if σ is the identity map on $[n+1]$.
On the other hand, suppose $\sigma(k) = n+1$ for some $k \in [n]$. We define a bijection τ based on σ by swapping the outputs corresponding to the inputs k and $n+1$. That is, we define $\tau : [n] \to [n]$ by

$$\tau(i) = \begin{cases} \sigma(i) & \text{if } i \neq k, n+1 \\ \sigma(n+1) & \text{if } i = k \\ \sigma(k) = n+1 & \text{if } i = n+1 \end{cases}.$$

By working backwards, we can show that it is true that

$$\sum_{i=1}^{n+1} x_i y_{\sigma(i)} \leq \sum_{i=1}^{n+1} x_i y_{\tau(i)}$$

because cancelling equal terms from both sides shows that it is equivalent to

$$\begin{aligned} x_k y_{\sigma(k)} + x_{n+1} y_{\sigma(n+1)} &\leq x_k y_{\tau(k)} + x_{n+1} y_{\tau(n+1)} \\ 0 &\leq x_k(y_{\tau(k)} - y_{\sigma(k)}) + x_{n+1}(y_{\tau(n+1)} - y_{\sigma(n+1)}) \\ &= x_k(y_{\sigma(n+1)} - y_{n+1}) + x_{n+1}(y_{n+1} - y_{\sigma(n+1)}) \\ &= (x_{n+1} - x_k)(y_{n+1} - y_{\sigma(n+1)}). \end{aligned}$$

This is true because $x_{n+1} \geq x_k$ and $y_{n+1} \geq y_{\sigma(n+1)}$. Then

$$\begin{aligned} \sum_{i=1}^{n+1} x_i y_{\sigma(i)} &\leq \sum_{i=1}^{n+1} x_i y_{\tau(i)} \\ &= \sum_{i=1}^{n} x_i y_{\tau(i)} + x_{n+1} y_{\tau(n+1)} = \sum_{i=1}^{n} x_i y_{\tau(i)} + x_{n+1} y_{n+1} \\ &\leq \sum_{i=1}^{n} x_i y_i + x_{n+1} y_{n+1} = \sum_{i=1}^{n+1} x_i y_i, \end{aligned}$$

where we used the induction hypothesis in the penultimate step. Thus, the rearrangement inequality holds in the case of $\sigma(n+1) \neq n+1$. For the sufficient equality condition, suppose

$$(x_1, x_2, \ldots, x_n, x_{n+1}) \text{ and } (y_1, y_2, \ldots, y_n, y_{n+1})$$

are strictly increasing. Then the strict inequality

$$(x_{n+1} - x_k)(y_{n+1} - y_{\sigma(n+1)}) > 0$$

holds because $x_{n+1} > x_k$ due to $k < n+1$ and $y_{n+1} > y_{\sigma(n+1)}$ due to $\sigma(n+1) \neq n+1$. So equality cannot hold for the overarching inequality because equality is impossible in this intermediate step. This completes the induction.

Finally, we will work on the left inequality

$$\sum_{i=1}^{n} x_i y_{n-i+1} \leq \sum_{i=1}^{n} x_i y_{\sigma(i)}.$$

Since $y_1 \leq y_2 \leq \cdots \leq y_n$, multiplying through by -1 yields $-y_n \leq -y_{n-1} \leq \cdots \leq -y_1$. By the right inequality that we painstakingly proved by induction, it holds for any bijection $\phi : [n] \to [n]$ that

$$\sum_{i=1}^{n} x_i(-y_{\phi(n-i+1)}) \leq \sum_{i=1}^{n} x_i(-y_{n-i+1}),$$

which is equivalent to

$$\sum_{i=1}^{n} x_i y_{n-i+1} \leq \sum_{i=1}^{n} x_i y_{\phi(n-i+1)}.$$

For any bijection $\sigma : [n] \to [n]$, define the bijection $\phi(i) = \sigma(n-i+1)$ for all i. Then $\sigma(i) = \phi(n-i+1)$; the intuition behind this is that ϕ and σ are reflections of each other. For this ϕ, the preceding inequality is

$$\sum_{i=1}^{n} x_i y_{n-i+1} \leq \sum_{i=1}^{n} x_i y_{\sigma(i)},$$

which is what we wanted to prove. Again, for the sufficient equality condition, suppose

$$(x_1, x_2, \ldots, x_n) \text{ and } (y_1, y_2, \ldots, y_n)$$

are strictly increasing. Then $(x_1, x_2, \ldots, x_n)$ and the reflected n-tuple $(-y_n, -y_{n-1}, \ldots, -y_1)$ are strictly increasing. Equality holds for the left side of the rearrangement inequality if and only if

$$\sum_{i=1}^{n} x_i y_{n-i+1} = \sum_{i=1}^{n} x_i y_{\sigma(i)},$$
$$\sum_{i=1}^{n} x_i(-y_{\sigma(i)}) = \sum_{i=1}^{n} x_i(-y_{n-i+1}),$$
$$\sum_{i=1}^{n} x_i(-y_{\phi(n-i+1)}) = \sum_{i=1}^{n} x_i(-y_{n-i+1}).$$

By the equality condition for the right side of the rearrangement inequality, equality holds in the last inequality above if and only if $\phi(n-i+1) = n-i+1$ for all $i \in [n]$, which is equivalent to it being true that $\sigma(i) = n-i+1$ for all i.

The other sufficient equality criterion, where at least one of the tuples

$$(x_1, x_2, \ldots, x_n) \text{ or } (y_1, y_2, \ldots, y_n)$$

is constant, is obvious, and we leave its two-line verification to the reader. □

Corollary 12.19 (Chebyshev's inequality)**.** Suppose n is a positive integer and

$$(x_1, x_2, \ldots, x_n),$$
$$(y_1, y_2, \ldots, y_n)$$

are each non-decreasing n-tuples of real numbers. Then

$$\frac{1}{n} \cdot \sum_{i=1}^{n} x_i y_i \geq \left(\frac{1}{n} \cdot \sum_{i=1}^{n} x_i\right)\left(\frac{1}{n} \cdot \sum_{j=1}^{n} y_j\right) \geq \frac{1}{n} \cdot \sum_{i=1}^{n} x_i y_{n-i+1}.$$

A case in which equality holds is when at least one of the n-tuples consists of all equal numbers.

Proof. By applying the rearrangement inequality to the n cyclically defined bijections

$$\begin{aligned}
\sigma_1 &: (1, 2, 3, \ldots, n-1, n) \mapsto (1, 2, 3, \ldots, n-2, n-1, n), \\
\sigma_2 &: (1, 2, 3, \ldots, n-1, n) \mapsto (2, 3, 4, \ldots, n-1, n, 1), \\
\sigma_3 &: (1, 2, 3, \ldots, n-1, n) \mapsto (3, 4, 5, \ldots, n, 1, 2), \\
&\vdots \\
\sigma_n &: (1, 2, 3, \ldots, n-1, n) \mapsto (n, 1, 2, \ldots, n-3, n-2, n-1),
\end{aligned}$$

and adding up the rearrangement inequalities, we get

$$n \cdot \sum_{i=1}^{n} x_i y_i \geq \sum_{j=1}^{n} \sum_{i=1}^{n} x_i y_{\sigma_j(i)} \geq n \cdot \sum_{i=1}^{n} x_i y_{n-i+1}.$$

Using the discrete Fubini's principle, the middle term can be rewritten as

$$\begin{aligned}
\sum_{j=1}^{n} \sum_{i=1}^{n} x_i y_{\sigma_j(i)} &= \sum_{i=1}^{n} \sum_{j=1}^{n} x_i y_{\sigma_j(i)} = \sum_{i=1}^{n} \left(x_i \sum_{j=1}^{n} y_{\sigma_j(i)} \right) \\
&= \sum_{i=1}^{n} \left(x_i \sum_{j=1}^{n} y_j \right) = \left(\sum_{i=1}^{n} x_i \right) \left(\sum_{j=1}^{n} y_j \right).
\end{aligned}$$

If $x_1 = x_2 = \cdots = x_n$, then equality clearly holds, and similarly for $y_1 = y_2 = \cdots = y_n$. □

12.4 Convexity

Definition 12.20. Let I be an interval (which is not necessarily bounded) in $\mathbb{R}$ and $f : I \to \mathbb{R}$ be a function. Then f is a **convex function** if, for all $x, y \in I$ such that $x \neq y$ and for all $\lambda \in (0, 1)$,

$$f((1-\lambda)x + \lambda y) \leq (1-\lambda)f(x) + \lambda f(y).$$

Note that this inequality automatically holds for $\lambda = 0$ and $\lambda = 1$ (try it out!), which is why it would be redundant to include them in the definition. If the inequality holds strictly for all $x \neq y$ in I and all $\lambda \in (0, 1)$, then f is said to be a **strictly convex function**. If the direction of the inequality is reversed, then we use the corresponding terms **concave** or **strictly concave**. In terms of graphs, convexity means that the line segment joining $(x, f(x))$ to $(y, f(y))$ does not dip below the graph of f from x to y, whereas strict convexity means the segment, excluding the endpoints, lies strictly above the graph of f.

Example. The second derivative of a function (meaning, the derivative of the derivative) is useful in determining its convexity, but we will not use this technique from calculus. Instead, we provide a list of functions that are well-known to be convex:

- x^n on $[0, \infty)$ for odd positive integers n

- x^n on $\mathbb{R}$ for even positive integers n
- b^x on $\mathbb{R}$ for real b such that $0 < b \neq 1$
- $\frac{1}{x^n}$ on $(0, \infty)$ for all positive integers n

Theorem 12.21 (Jensen's inequality). Let I be a real interval, $f : I \to \mathbb{R}$ be a convex function, n be a positive integer, $x_1, x_2, \ldots, x_n \in I$, and $\lambda_1, \lambda_2, \ldots, \lambda_n \in [0, 1]$ such that

$$\lambda_1 + \lambda_2 + \cdots + \lambda_n = 1.$$

Then

$$f(\lambda_1 x_1 + \lambda_2 x_2 + \cdots + \lambda_n x_n) \leq \lambda_1 f(x_1) + \lambda_2 f(x_2) + \cdots + \lambda_n f(x_n).$$

If f is strictly convex, then equality holds if and only if all x_i with non-zero corresponding λ_i are equal. A special case of the inequality occurs at

$$\lambda_1 = \lambda_2 = \cdots = \lambda_n = \frac{1}{n},$$

where the inequality becomes

$$f\left(\frac{x_1 + x_2 + \cdots + x_n}{n}\right) \leq \frac{f(x_1) + f(x_2) + \cdots + f(x_n)}{n}.$$

Proof. We proceed by induction on n. For $n = 1$, $\lambda_1 = 1$ and Jensen says

$$f(\lambda_1 x_1) \leq \lambda_1 f(x_1),$$

which is true with equality. For $n = 2$, Jensen says

$$f(\lambda_1 x_1 + \lambda_2 x_2) \leq \lambda_1 f(x_1) + \lambda_2 f(x_2).$$

This is true by the convexity of f because $\lambda_1 = 1 - \lambda_2$.
Now suppose the result holds for some $n - 1 \in \mathbb{Z}_+$. As the inductive step, suppose we are working in the successive case of n numbers. Our strategy will be to separate $\lambda_n x_n$ from the rest of the terms so that the convexity of f can be used, followed by the induction hypothesis on the remaining $n - 1$ terms. If $\lambda_n = 1$, then all other λ_i are 0, which reduces the scenario to the $n = 1$ case. So, suppose $\lambda_n \neq 1$. First we perform the manipulation

$$f(\lambda_1 x_1 + \lambda_2 x_2 + \cdots + \lambda_n x_n) = f\left[(1 - \lambda_n)\left(\frac{\lambda_1}{1 - \lambda_n} \cdot x_1 + \cdots + \frac{\lambda_{n-1}}{1 - \lambda_n} \cdot x_{n-1}\right) + \lambda_n x_n\right].$$

Since $(1 - \lambda_n) + \lambda_n = 1$, the convexity of f says that an upper bound on the above expression is

$$(1 - \lambda_n) f\left(\frac{\lambda_1}{1 - \lambda_n} \cdot x_1 + \frac{\lambda_2}{1 - \lambda_n} \cdot x_2 + \cdots + \frac{\lambda_{n-1}}{1 - \lambda_n} \cdot x_{n-1}\right) + \lambda_n f(x_n).$$

Now we wish to use the induction hypothesis on the first term. Note that

$$\sum_{i=1}^{n-1} \frac{\lambda_i}{1 - \lambda_n} = \frac{\lambda_1 + \lambda_2 + \cdots + \lambda_{n-1}}{1 - \lambda_n} = \frac{1 - \lambda_n}{1 - \lambda_n} = 1.$$

Rearranging the indices if needed, we may assume that x_1 is the smallest among the x_i and that x_n is the largest among the x_i. Then

$$x_1 = x_1 \cdot \sum_{i=1}^{n-1} \frac{\lambda_i}{1 - \lambda_n} \leq \sum_{i=1}^{n-1} \left(\frac{\lambda_i}{1 - \lambda_n} \cdot x_i\right) \leq x_n \cdot \sum_{i=1}^{n-1} \frac{\lambda_i}{1 - \lambda_n} = x_n,$$

so the central sum actually lies in the original interval I. By the induction hypothesis,

$$\begin{aligned}(1-\lambda_n)f\left[\sum_{i=1}^{n-1}\left(\frac{\lambda_i}{1-\lambda_n}\cdot x_i\right)\right]+\lambda_n f(x_n) &\le (1-\lambda_n)\sum_{i=1}^{n-1}\left(\frac{\lambda_i}{1-\lambda_n}f(x_i)\right)+\lambda_n f(x_n)\\ &= \lambda_1 f(x_1)+\lambda_2 f(x_2)+\cdots+\lambda_n f(x_n).\end{aligned}$$

This proves Jensen's inequality. Now we tackle the equality condition. Supposing f is strictly convex, we will make some reductions:

- If some λ_k is 1, then all other λ_i are 0. Then the condition that "all x_i, for which λ_i is non-zero, are equal" becomes trivial. So, we will suppose in the below cases that no λ_k is 1.
- If we can prove the equality criterion for all $\lambda_1, \lambda_2, \ldots, \lambda_n \in (0,1)$, it will imply the case where some λ_k are 0 because those $\lambda_k x_k$ and $\lambda_k f(x_k)$ terms disappear from the inequality, leading to a case where no λ_i is 0. So suppose no λ_k is 0 in the final case below.
- As stated in the first and second bullets, we can now assume that all λ_i are in $(0,1)$. We break this case into several steps:
  - In one direction, if all the x_i are equal, then it is not difficult to verify that equality holds in Jensen. Below, we tackle the other direction.
  - If all the x_i are distinct, then we proceed by induction to show that Jensen is strict. In the base case of $n=1$, since f is strictly convex, the definition of strict convexity requires that, if $x_1 \neq x_2$ and λ_1, λ_2 are non-zero, then the inequality is strict. Assuming the induction hypothesis for $x_1 < x_2 < \cdots < x_n$ for some $n \geq 2$, it is proven for $x_1 < x_2 < \cdots < x_n < x_{n+1}$ by noting that the application of the inductive hypothesis at the end of the proof of the original Jensen's inequality makes the inequality strict.
  - Now suppose the x_i are neither all distinct nor all equal for some $n > 2$ (n must be strictly greater than 2 because two numbers are either the same as each other or different). Then the terms of Jensen, which says

    $$f\left(\sum_{i=1}^{n}\lambda_i x_i\right) \le \sum_{i=1}^{n}\lambda_i f(x_i),$$

    can be regrouped by collecting like terms as

    $$f\left(\sum_{i=1}^{m}\ell_i y_i\right) \le \sum_{i=1}^{m}\ell_i f(y_i),$$

    where the y_i are distinct and $m < n$. By the previous case, this must be strict.

□

Problem 12.22 (Power means inequality ladder). Let $n \in \mathbb{Z}_+$, $(x_i)_{i=1}^n$ be a list of positive reals, and $(\lambda_i)_{i=1}^n$ be a list of positive reals such that

$$\lambda_1+\lambda_2+\cdots+\lambda_n=1.$$

Prove that, if $r > s$ are non-zero real numbers, then

$$(\lambda_1 x_1^r+\lambda_2 x_2^r+\cdots+\lambda_n x_n^r)^{\frac{1}{r}} \ge (\lambda_1 x_1^s+\lambda_2 x_2^s+\cdots+\lambda_n x_n^s)^{\frac{1}{s}}.$$

Also prove that equality holds if and only if $x_1=x_2=\cdots=x_n$.

Theorem 12.23 (Weighted AM-GM inequality). Let $n \in \mathbb{Z}_+$, $(x_i)_{i=1}^n$ be a list of non-negative reals, and $(\lambda_i)_{i=1}^n$ be a list of non-negative reals such that $\lambda_1+\lambda_2+\cdots+\lambda_n=1$. Then

$$\lambda_1 x_1+\lambda_2 x_2+\cdots+\lambda_n x_n \ge x_1^{\lambda_1}x_2^{\lambda_2}\cdots x_n^{\lambda_n},$$

where we use the temporary convention that $0^0 = 1$. Equality holds if and only if all x_i, such that $\lambda_i \neq 0$, are equal. We can lift the criterion that the λ_i add to 1, and instead set $\lambda_1 + \lambda_2 + \cdots + \lambda_n = \lambda$ to get the corollary

$$\frac{\lambda_1 x_1 + \lambda_2 x_2 + \cdots + \lambda_n x_n}{\lambda} \geq (x_1^{\lambda_1} x_2^{\lambda_2} \cdots x_n^{\lambda_n})^{\frac{1}{\lambda}}.$$

Setting all $\lambda_i = 1$ in the corollary yields the ordinary AM-GM inequality.

Proof. We assume, without loss of generality, that all x_i are positive, as any x_i being 0 reduces the inequality to a case with fewer variables and where all remaining x_i are positive. The key is to use Jensen's inequality with the strictly convex function $f(x) = e^x$. It is helpful to observe that

$$e^{t \cdot \ln x} = x^t$$

holds for all positive real x and all real t. By Jensen,

$$\begin{aligned} x_1^{\lambda_1} x_2^{\lambda_2} \cdots x_n^{\lambda_n} &= e^{\lambda_1 \ln x_1} e^{\lambda_2 \ln x_2} \cdots e^{\lambda_n \ln x_n} \\ &= e^{\lambda_1 \ln x_1 + \lambda_2 \ln x_2 + \cdots + \lambda_n \ln x_n} \\ &\leq \lambda_1 e^{\ln x_1} + \lambda_2 e^{\ln x_2} + \cdots + \lambda_n e^{\ln x_n} \\ &= \lambda_1 x_1 + \lambda_2 x_2 + \cdots + \lambda_n x_n. \end{aligned}$$

The equality condition for Jensen's inequality says that equality holds if and only if all of the $\ln x_i$, for which $\lambda_i \neq 0$, are equal. By inverting the logarithm, this condition is true if and only if all of the x_i, for which $\lambda_i \neq 0$, are equal. The corollaries are immediately true. □

Problem 12.24 (Generalized Young's inequality). Let $n \in \mathbb{Z}_+$, $(x_i)_{i=1}^n$ be a list of non-negative reals, and $(\mu_i)_{i=1}^n$ be a list of positive reals, each greater than 1, such that

$$\frac{1}{\mu_1} + \frac{1}{\mu_2} + \cdots + \frac{1}{\mu_n} = 1.$$

Prove that

$$x_1 x_2 \cdots x_n \leq \frac{x_1^{\mu_1}}{\mu_1} + \frac{x_2^{\mu_2}}{\mu_2} + \cdots + \frac{x_n^{\mu_n}}{\mu_n},$$

with equality holding if and only if

$$x_1^{\mu_1} = x_2^{\mu_2} = \cdots = x_n^{\mu_n}.$$

Theorem 12.25 (Hölder's inequality). Let $n, m \in \mathbb{Z}_+$, and $(x_{1,i})_{i=1}^n, (x_{2,i})_{i=1}^n, \ldots, (x_{m,i})_{i=1}^n$ be m lists, each of n non-negative real numbers. Let $\lambda_1, \lambda_2, \ldots, \lambda_n$ be n non-negative real numbers that satisfy $\lambda_1 + \lambda_2 + \cdots + \lambda_n = 1$. Then

$$\prod_{j=1}^{m} \left(\sum_{i=1}^{n} x_{j,i} \right)^{\lambda_j} \geq \sum_{i=1}^{n} \left(\prod_{j=1}^{m} x_{j,i}^{\lambda_j} \right),$$

with the convention that $0^0 = 1$. This is written in more leisurely notation as

$$\begin{aligned} &(x_{1,1} + x_{1,2} + \cdots + x_{1,n})^{\lambda_1} \cdot (x_{2,1} + x_{2,2} + \cdots + x_{2,n})^{\lambda_2} \cdots (x_{m,1} + x_{m,2} + \cdots + x_{m,n})^{\lambda_m} \\ &\qquad \geq (x_{1,1}^{\lambda_1} x_{2,1}^{\lambda_2} \cdots x_{m,1}^{\lambda_m}) + (x_{1,2}^{\lambda_1} x_{2,2}^{\lambda_2} \cdots x_{m,2}^{\lambda_m}) + \cdots + (x_{1,n}^{\lambda_1} x_{2,n}^{\lambda_2} \cdots x_{m,n}^{\lambda_m}). \end{aligned}$$

Visually, we can place the $x_{j,i}$ in an $m \times n$ matrix with each list occupying a row, so that the inequality's left side has the sum of a row in each multiplicand. The inequality's right side similarly has the product of a column in each summand, but with a matrix of $x_{j,i}^{\lambda_j}$'s as entries. Equality holds if and only if, for all $j \in [m]$ such that $\lambda_j \neq 0$,

$$x_{j,1} : x_{j,2} : \cdots : x_{j,m}$$

forms the same ratio, or one of the lists $(x_{j,i})_{i=1}^n$ consists of all 0's.

Proof. If some $x_{j,i}$ is 0, then the desired inequality is weaker than (and, so, it is implied by) Hölder with the entire i^{th} column of the matrix being 0; many terms disappear in the latter to produce a Hölder's inequality with $(n-1)$-length lists. This can be repeated for any other zero terms. If any list consists of all 0's, then the inequality is trivially true as an equality. Thus, we may assume, without loss of generality, that all $x_{j,i}$ are positive. Dividing both sides of the inequality by the left side yields

$$1 \geq \sum_{i=1}^{n} \prod_{j=1}^{m} \left(\frac{x_{j,i}}{\sum_{k=1}^{n} x_{j,k}} \right)^{\lambda_j}.$$

For each $j \in [m]$ and $i \in [n]$, let

$$y_{j,i} = \frac{x_{j,i}}{\sum_{k=1}^{m} x_{j,k}},$$

so that we obtain the condition

$$y_{j,1} + y_{j,2} + \cdots + y_{j,n} = 1$$

for each $j \in [m]$. We wish to prove that

$$\sum_{i=1}^{n} \prod_{j=1}^{m} y_{j,i}^{\lambda_j} \leq 1.$$

By applying the weighted AM-GM inequality n times and the discrete Fubini's principle,

$$\begin{aligned} \sum_{i=1}^{n} \prod_{j=1}^{m} y_{j,i}^{\lambda_j} \leq \sum_{i=1}^{n} \sum_{j=1}^{m} \lambda_j y_{j,i} &= \sum_{j=1}^{m} \sum_{i=1}^{n} \lambda_j y_{j,i} \\ &= \sum_{j=1}^{m} \left(\lambda_j \sum_{i=1}^{n} y_{j,i} \right) \\ &= \sum_{j=1}^{m} \lambda_j = 1, \end{aligned}$$

which establishes the inequality.

Now we need to work on the equality condition. By the equality condition for the weighted AM-GM inequality, equality holds if and only if, for each fixed $i \in [n]$ and for all $j \in [m]$ such that $\lambda_j \neq 0$, the $y_{j,i}$ are equal. Taking all λ_j to be non-zero for ease of notation, this means that there exists a positive real t_i such that

$$t_i = y_{1,i} = y_{2,i} = \cdots = y_{m,i}.$$

For each $j \in [m]$, let

$$s_j = x_{j,1} + x_{j,2} + \cdots + x_{j,n} \neq 0.$$

Then, for each $i \in [n]$ and $j \in [m]$,

$$t_i = y_{j,i} = \frac{x_{j,i}}{s_j} \implies x_{j,i} = s_j t_i.$$

Fixing $j \in [m]$, we get

$$x_{j,1} : x_{j,2} : \cdots : x_{j,n} = s_j t_1 : s_j t_2 : \cdots : s_j t_n = t_1 : t_2 : \cdots : t_n,$$

which is the same for every $j \in [m]$. The converse, that the criterion implies equality, is easy to establish algebraically. If one of the lists is all 0's, equality automatically holds; we addressed this possibility at the beginning of the proof. □

Problem 12.26. Let $n \in \mathbb{Z}_+$, let $(x_k)_{k=1}^n$ and $(y_k)_{k=1}^n$ each be a list of n non-negative real numbers, and let p and q be real numbers greater than 1 such that $\frac{1}{p} + \frac{1}{q} = 1$. As a special case of Hölder's inequality, prove that

$$\sum_{k=1}^{n} x_k y_k \leq \left(\sum_{k=1}^{n} x_k^p \right)^{\frac{1}{p}} \cdot \left(\sum_{k=1}^{n} x_k^q \right)^{\frac{1}{q}}.$$

Also, prove that equality holds if and only if the list of x_i's is all 0's, or the list of y_i's is all 0's, or there exists a positive real c such that, for all $k \in [n]$, $y_k^q = cx_k^p$.

Theorem 12.27 (Minkowski's inequality)**.** Let $n \in \mathbb{Z}_+$ and $p \geq 1$ be a real number. The p**-norm** of an n-tuple $\mathbf{x} = (x_1, x_2, \dots, x_n)$ of real numbers is denoted by and defined as

$$\|\mathbf{x}\|_p = (|x_1|^p + |x_2|^p + \cdots + |x_n|^p)^{\frac{1}{p}}.$$

Minkowski says that, for any two n-tuples $\mathbf{x}, \mathbf{y} \in \mathbb{R}^n$ and any real $p \geq 1$,

$$\|\mathbf{x} + \mathbf{y}\|_p \leq \|\mathbf{x}\|_p + \|\mathbf{y}\|_p.$$

This can be considered to be the triangle inequality on the p-norm. Equality holds if and only if all the entries of $\mathbf{x}$ are 0 or all the entries of $\mathbf{y}$ are 0, or there exists a positive real c such that $\mathbf{y} = c\mathbf{x}$.

Proof. The $p = 1$ case follows from the usual triangle inequality, so we will assume $p > 1$. Then $\dfrac{1}{p} < 1$, and we let $q > 1$ be the real number such that $\dfrac{1}{p} + \dfrac{1}{q} = 1$. By the real triangle inequality (Theorem 6.4),

$$\begin{aligned}
\|\mathbf{x} + \mathbf{y}\|_p^p &= \sum_{k=1}^{n} |x_k + y_k|^p \\
&= \sum_{k=1}^{n} |x_k + y_k|^{p-1} \cdot |x_k + y_k| \\
&\leq \sum_{k=1}^{n} |x_k + y_k|^{p-1} \cdot (|x_k| + |y_k|) \\
&= \sum_{k=1}^{n} |x_k + y_k|^{p-1} \cdot |x_k| + \sum_{k=1}^{n} |x_k + y_k|^{p-1} \cdot |y_k|.
\end{aligned}$$

So far, by the criterion for equality in the real triangle inequality, equality holds if and only if, for each $k \in [n]$, x_k of y_k are not on opposite sides of 0 on the real number line, meaning there exists a real constant $c_k \geq 0$ such that $y_k = c_k x_k$. Moving forward with proving the inequality, we will use the equivalences

$$\frac{1}{p} + \frac{1}{q} = 1 \iff p + q = pq \iff p = (p-1)q \iff 1 = p - \frac{p}{q}.$$

By Hölder's inequality, we continue with

$$\begin{aligned}
\|\mathbf{x} + \mathbf{y}\|_p^p &\leq \sum_{k=1}^{n} |x_k + y_k|^{p-1} \cdot |x_k| + \sum_{k=1}^{n} |x_k + y_k|^{p-1} \cdot |y_k| \\
&\leq \left(\sum_{k=1}^{n} |x_k + y_k|^{(p-1)q}\right)^{\frac{1}{q}} \left(\sum_{k=1}^{n} |x_k|^p\right)^{\frac{1}{p}} + \left(\sum_{k=1}^{n} |x_k + y_k|^{(p-1)q}\right)^{\frac{1}{q}} \left(\sum_{k=1}^{n} |y_k|^p\right)^{\frac{1}{p}} \\
&= \left(\sum_{k=1}^{n} |x_k + y_k|^{p}\right)^{\frac{1}{q}} \left(\sum_{k=1}^{n} |x_k|^p\right)^{\frac{1}{p}} + \left(\sum_{k=1}^{n} |x_k + y_k|^{p}\right)^{\frac{1}{q}} \left(\sum_{k=1}^{n} |y_k|^p\right)^{\frac{1}{p}} \\
&= \|\mathbf{x} + \mathbf{y}\|_p^{\frac{p}{q}} \cdot (\|\mathbf{x}\|_p + \|\mathbf{y}\|_p).
\end{aligned}$$

As a result, the inequality is established as

$$\|\mathbf{x} + \mathbf{y}\|_p = \|\mathbf{x} + \mathbf{y}\|_p^{p - \frac{p}{q}} \leq \|\mathbf{x}\|_p + \|\mathbf{y}\|_p.$$

Now we complete the proof of the equality criteria. If $\mathbf{x} = 0$ or $\mathbf{y} = 0$, equality can be verified to hold, so suppose neither list consists of all 0's. Equality holds in the second part of the proof if and only if equality holds in the application of Hölder's inequality. By the equality condition on Hölder, equality holds if and only if there exist non-negative real constants d_1, d_2 such that, for all $k \in [n]$,

$$d_1|x_k|^p = |x_k + y_k|^{(p-1)q} = |x_k + y_k|^p,$$
$$d_2|y_k|^p = |x_k + y_k|^{(p-1)q} = |x_k + y_k|^p.$$

Then $d_1|x_k|^p = d_2|y_k|^p$ for all $k \in [n]$. Since $\mathbf{x}, \mathbf{y}$ are non-zero lists, $d_1 \neq 0$ and $d_2 \neq 0$. So, letting $c = \left(\frac{d_1}{d_2}\right)^{\frac{1}{p}} > 0$, we get $|y_k| = c|x_k|$ for all $k \in [n]$. The equality condition from the first half of the proof requires that x_k, y_k are not of opposite signs, so we can remove the absolute values to get $\mathbf{x} = c\mathbf{y}$. Conversely, the fact that equality holds under the stated condition is easy to establish algebraically. □

Theorem 12.28 (Popoviciu's inequality). Let I be a real interval, and $f : I \to \mathbb{R}$ be a convex function. Then, for any $a, b, c \in I$,

$$\frac{2}{3} \cdot \left[f\left(\frac{a+b}{2}\right) + f\left(\frac{b+c}{2}\right) + f\left(\frac{c+a}{2}\right)\right] \leq \frac{f(a) + f(b) + f(c)}{3} + f\left(\frac{a+b+c}{3}\right).$$

Proof. Due to a, b, c occupying symmetric positions, we may assume, without loss of generality, that $a \leq b \leq c$. We will split the proof into cases according to how the middle variable b compares to the average $\frac{a+b+c}{3}$. There are two cases:

1. Suppose $b \leq \frac{a+b+c}{3}$. It may be verified that

$$\frac{a+b+c}{3} \leq \frac{a+c}{2} \leq c,$$
$$\frac{a+b+c}{3} \leq \frac{b+c}{2} \leq c.$$

Then there exist $\lambda_a, \lambda_b \in [0, 1]$ such that

$$\frac{a+c}{2} = (1 - \lambda_a) \cdot \frac{a+b+c}{3} + \lambda_a c,$$
$$\frac{b+c}{2} = (1 - \lambda_b) \cdot \frac{a+b+c}{3} + \lambda_b c.$$

The sum of these equations is

$$\begin{aligned}\frac{a+b+2c}{2} &= (2 - \lambda_a - \lambda_b) \cdot \frac{a+b+c}{3} + (\lambda_a + \lambda_b)c \\ &= (2 - \lambda_a - \lambda_b) \cdot \frac{a+b-2c}{3} + 2c,\end{aligned}$$

which can be simplified to

$$\frac{a+b-2c}{2} = (2 - \lambda_a - \lambda_b) \cdot \frac{a+b-2c}{3}.$$

If $a + b - 2c = 0$, then the fact that $a \leq c$ and $b \leq c$ leads to $a = b = c$, in which case Popoviciu holds. Supposing $a + b \neq 2c$, we can cancel it from both sides to get

$$\frac{1}{2} = \frac{2 - \lambda_a - \lambda_b}{3} \implies \lambda_a + \lambda_b = \frac{1}{2}.$$

Using earlier equations, we work on developing upper bounds for the components on the left side of the desired inequality. Due to the convexity of f, Jensen says that

$$f\left(\frac{a+b}{2}\right) \leq \frac{f(a) + f(b)}{2},$$
$$f\left(\frac{b+c}{2}\right) \leq (1 - \lambda_a) f\left(\frac{a+b+c}{3}\right) + \lambda_a f(c),$$
$$f\left(\frac{c+a}{2}\right) \leq (1 - \lambda_b) f\left(\frac{a+b+c}{3}\right) + \lambda_b f(c).$$

Adding the three and dividing both sides by $\frac{3}{2}$ yields Popoviciu's inequality, since $\lambda_a + \lambda_b = \frac{1}{2}$.

2. Suppose $b \geq \dfrac{a+b+c}{3}$. Then we proceed as in the first case, but with the inequalities

$$a \leq \frac{a+c}{2} \leq \frac{a+b+c}{3},$$
$$a \leq \frac{a+b}{2} \leq \frac{a+b+c}{3}.$$

The reader is strongly encouraged to write the proof of this case by mimicking the technicalities of the first case.

□

A lesser-known inequality involving convexity is that of Karamata. A proof is easily found in other sources.

12.5 Newton and Maclaurin

Definition 12.29. Let n be a positive integer, k be an integer such that $1 \leq k \leq n$ and let $r_1, r_2, \ldots, r_n$ be a list of real numbers. Recall from Definition 11.44 that the kth symmetric sum of this list is

$$\sigma_k = \sum_{\substack{J \subseteq [n] \\ |J|=k}} \prod_{j \in J} r_j = \sum_{1 \leq i_1 < i_2 < \cdots < i_k \leq n} r_{i_1} r_{i_2} \cdots r_{i_k}.$$

Now we define the kth **symmetric mean** as the average of the terms of the kth symmetric sum. Since the number of k-tuples $(i_1, i_2, \ldots, i_n)$ such that

$$1 \leq i_1 < i_2 < \cdots < i_k \leq n$$

is $\dbinom{n}{k}$ by a bijective counting argument from Volume 2, it means the kth symmetric mean of the list $r_1, r_2, \ldots, r_n$ is

$$\mu_k = \frac{\sigma_k}{\binom{n}{k}}.$$

We also extend the definitions to $k = 0$ to get the empty sum $\sigma_0 = 1$ and $\mu_0 = \dfrac{\sigma_0}{\binom{n}{0}} = 1$.

Theorem 12.30 (Newton's inequalities). If $n \geq 2$ is an integer, $r_1, r_2, \ldots, r_n$ are non-negative real numbers, and k is an integer such that $0 < k < n$, then

$$\mu_k^2 \geq \mu_{k-1}\mu_{k+1}.$$

If all r_i are positive, then equality holds for a specific k if and only if all the r_i are equal to each other. As a result, equality holds for a specific k if and only if it holds for all k.

Proof. We will prove this by induction on $n \geq 2$, handling all k at each step after fixing n. In the base case of $n = 2$, the only possibility for k is $k = 1$, and the inequality says

$$\left(\frac{r_1 + r_2}{2}\right)^2 \geq 1 \cdot (r_1 r_2).$$

As we know, this is the most basic case of AM-GM, with equality holding if and only if $r_1 = r_2$.
Now suppose the result holds for any $n-1$ non-negative real numbers for some integer n such that $n - 1 \geq 2$. To prove Newton's inequality for n numbers $r_1, r_2, \ldots, r_n$, we will first set $(\nu_k)_{k=1}^{n-1}$ to be the symmetric means of $r_1, r_2, \ldots, r_{n-1}$ (specifically, r_n is excluded). Then, for each integer k such that $0 < k < n$, we split μ_k

into the terms that do not contain r_n and the terms that do contain r_n. We then utilize the combinatorial identities (see Volume 2)

$$\binom{n}{k} = \frac{n}{k} \cdot \binom{n-1}{k-1},$$
$$\binom{n}{k} = \binom{n}{n-k} = \frac{n}{n-k} \cdot \binom{n-1}{n-k-1} = \frac{n}{n-k} \cdot \binom{n-1}{k}$$

to get the recursion

$$\mu_k = \frac{n-k}{n} \cdot \nu_k + \frac{k}{n} \cdot \nu_{k-1} r_n.$$

Convince yourself that this is true. The preceding recursion is the only lemma that we will need.

If we wish to use the above recursion on μ_{k-1} in $\mu_{k-1}\mu_{k+1}$, it will result in a term with ν_{k-2}, which is sensible only for $k \geq 2$. So, we will handle $k = 1$ separately at the outset now. If $k = 1$, then Newton says

$$\begin{aligned} \left(\frac{r_1 + r_2 + \cdots + r_n}{n}\right)^2 &\geq 1 \cdot \left(\sum_{1 \leq i < j \leq n} r_i r_j\right) \cdot \frac{1}{\binom{n}{2}} \\ &= \frac{2}{n(n-1)} \cdot \sum_{1 \leq i < j \leq n} r_i r_j. \end{aligned}$$

Further backwards manipulations yield

$$\begin{aligned} (n-1)\left[\sum_{k=1}^{n} r_k^2 + 2 \sum_{1 \leq i < j \leq n} r_i r_j\right] &\geq 2n \sum_{1 \leq i < j \leq n} r_i r_j \\ (n-1)\sum_{k=1}^{n} r_k^2 &\geq 2 \sum_{1 \leq i < j \leq n} r_i r_j \\ \sum_{1 \leq i < j \leq n} (r_i^2 + r_j^2) &\geq \sum_{1 \leq i < j \leq n} 2 r_i r_j \\ \sum_{1 \leq i < j \leq n} (r_i - r_j)^2 &\geq 0, \end{aligned}$$

which is true by the trivial inequality. Equality holds if and only if

$$r_1 = r_2 = \cdots = r_n.$$

Now we may assume that $k \geq 2$. By the recursive lemma,

$$\begin{aligned} \mu_{k-1}\mu_{k+1} &= \left(\frac{n-k+1}{n} \cdot \nu_{k-1} + \frac{k-1}{n} \cdot \nu_{k-2} r_n\right) \cdot \left(\frac{n-k-1}{n} \cdot \nu_{k+1} + \frac{k+1}{n} \cdot \nu_k r_n\right) \\ &= \frac{(n-k+1)(n-k-1)}{n^2} \cdot \nu_{k-1}\nu_{k+1} + \frac{(k-1)(n-k-1)}{n^2} \cdot \nu_{k-2}\nu_{k+1} r_n \\ &+ \frac{(n-k+1)(k+1)}{n^2} \cdot \nu_{k-1}\nu_k r_n + \frac{(k-1)(k+1)}{n^2} \cdot \nu_{k-2}\nu_k r_n^2. \end{aligned}$$

We use the induction hypothesis to get

$$\begin{aligned} \nu_{k-1}\nu_{k+1} &\leq \nu_k^2, \\ \nu_{k-2}\nu_k &\leq \nu_{k-1}^2, \end{aligned}$$

and multiplying them together yields

$$\nu_{k-2}\nu_{k-1}\nu_k\nu_{k+1} \leq \nu_{k-1}^2 \nu_k^2.$$

We wish to cancel $\nu_{k-1}\nu_k$ from both sides, but we need to handle the possibilities of $\nu_k = 0$ or $\nu_{k-1} = 0$ first.

- Suppose $\nu_k = 0$. Then the fact that $\nu_{k-1}\nu_{k+1} \leq \nu_k^2$ says that $\nu_{k-1} = 0$ or $\nu_{k+1} = 0$.
  - Suppose $\nu_{k-1} = 0$. Then the recursion says that $\mu_k = 0$, so the product of any collection of k of the r_i multiply to 0. Substituting 0 for ν_k, ν_{k-1}, μ_k into the above expansion of $\mu_{k-1}\mu_{k+1}$ and into Newton, we find that proving Newton is equivalent to proving
  $$0 = \mu_k^2 \geq \mu_{k-1}\mu_{k+1} = \frac{(k-1)(n-k-1)}{n^2} \cdot \nu_{k-2}\nu_{k+1}r_n \iff \nu_{k-2}\nu_{k+1}r_n = 0.$$
  The fact that $\nu_k = 0$ means the product of any k of $r_1, r_2, \ldots, r_{n-1}$ is 0, so any $k+1$ of them must also multiply to 0, which gives $\nu_{k+1} = 0$, thus proving Newton in this case.
  - Suppose $\nu_{k+1} = 0$. By the above expansion of $\mu_{k-1}\mu_{k+1}$, we get $\mu_{k-1}\mu_{k+1} = 0$, so $\mu_{k-1} = 0$ or $\mu_{k+1} = 0$, either of which makes Newton true by the trivial inequality.
- Suppose $\nu_{k-1} = 0$. Then the fact that $\nu_{k-2}\nu_k \leq \nu_{k-1}^2$ says that $\nu_{k-2} = 0$ or $\nu_k = 0$.
  - Suppose $\nu_{k-2} = 0$. Again, by the expansion of $\mu_{k-1}\mu_{k+1}$ above, $\mu_{k-1}\mu_{k+1} = 0$, which we have previously shown to imply Newton.
  - Suppose $\nu_k = 0$. We have shown above that the combination of $\nu_k = 0$ and $\nu_{k-1} = 0$ implies Newton.

So now we may assume that $\nu_{k-1}\nu_k \neq 0$ and cancel it from both sides of
$$\nu_{k-2}\nu_{k-1}\nu_k\nu_{k+1} \leq \nu_{k-1}^2\nu_k^2$$
to get
$$\nu_{k-2}\nu_{k+1} \leq \nu_{k-1}\nu_k.$$
Continuing with the upper bounding of $\mu_{k-1}\mu_{k+1}$, we get
$$\mu_{k-1}\mu_{k+1} \leq \frac{(n-k)^2-1}{n^2} \cdot \nu_k^2 + \frac{(k-1)(n-k-1)}{n^2} \cdot \nu_{k-1}\nu_k r_n + \frac{(n-k+1)(k+1)}{n^2} \cdot \nu_{k-1}\nu_k r_n + \frac{k^2-1}{n^2} \cdot \nu_{k-1}^2 r_n^2.$$
To collect the middle two like terms, expansion and simplification yield the coefficient
$$\frac{(k-1)(n-k-1)}{n^2} + \frac{(n-k+1)(k+1)}{n^2} = \frac{2k(n-k)}{n^2} + \frac{2}{n^2}.$$
Finally, we reach the desired upper bound:
$$\begin{aligned}\mu_{k-1}\mu_{k+1} &\leq \frac{(n-k)^2}{n^2} \cdot \nu_k^2 + \frac{2k(n-k)}{n^2} \cdot \nu_k\nu_{k-1}r_n + \frac{k^2}{n^2} \cdot \nu_{k-1}^2 r_n^2 - \frac{\nu_k^2}{n^2} + \frac{2}{n^2} \cdot \nu_k\nu_{k-1}r_n - \frac{\nu_{k-1}^2 r_n^2}{n^2} \\ &= \left(\frac{n-k}{n} \cdot \nu_k + \frac{k}{n} \cdot \nu_{k-1}r_n\right)^2 - \left(\frac{\nu_k}{n} - \frac{\nu_{k-1}r_n}{n}\right)^2 \leq \mu_k^2 - 0^2 = \mu_k^2.\end{aligned}$$
For the equality condition, suppose all of the r_i are positive. If equality holds, it must have held in both of
$$\nu_{k-1}\nu_{k+1} \leq \nu_k^2,$$
$$\left(\frac{\nu_k}{n} - \frac{\nu_{k-1}r_n}{n}\right)^2 \leq 0.$$
By the equality criterion in the induction hypothesis, if $\nu_{k-1}\nu_{k+1} = \nu_k^2$, then there exists a positive real r such that
$$r = r_1 = r_2 = \cdots = r_{n-1}.$$
By the second condition, which states $\nu_k = \nu_{k-1}r_n$, we can substitute $r > 0$ for each r_i in ν_k and ν_{k-1} to get
$$\frac{\binom{n-1}{k}r^k}{\binom{n-1}{k}} = \frac{\binom{n-1}{k-1}r^{k-1}}{\binom{n-1}{k-1}} r_n \implies r = r_n.$$
Conversely, it is not difficult to verify algebraically that, if all r_i are positive and equal, then equality holds in Newton's inequality. □

Problem 12.31. Let n be a positive integer. Prove that, for any $a_1, a_2, \ldots, a_n \in \mathbb{R}$, if

$$\forall k \in [n-1] : \frac{a_{k-1} + a_{k+1}}{2} \geq a_k,$$

then

$$\forall k \in [n-1] : \frac{a_{k+1} - a_0}{k+1} \geq \frac{a_k - a_0}{k}.$$

Theorem 12.32 (Maclaurin's inequalities). Let $n \geq 2$ be an integer, and $r_2, r_2, \ldots, r_n$ be positive real numbers. For each $k \in [n-1]$,

$$(\mu_k)^{\frac{1}{k}} \geq (\mu_{k+1})^{\frac{1}{k+1}},$$

where we have used the notation of symmetric means (Definition 12.29). Equality holds for a specific k if and only if $r_1 = r_2 = \cdots = r_n$. So, equality holds for a specific k if and only if it holds for all k. Altogether, the inequalities state that

$$\mu_1 \geq \sqrt{\mu_2} \geq \sqrt[3]{\mu_3} \geq \cdots \geq \sqrt[n]{\mu_n}.$$

Since μ_1 is the arithmetic mean of the r_i and $\sqrt[n]{\mu_n}$ is their geometric mean, Maclaurin's inequalities point out gradations or refinements in between the two sides of the AM-GM inequality.

Proof. By taking the natural logarithm of Newton's inequalities (which is acceptable, since all of the r_i are positive, so their symmetric sums are also positive, allow them to be fed into logarithms),

$$\begin{aligned} \mu_k &\geq (\mu_{k-1}\mu_{k+1})^{\frac{1}{2}} \\ \ln \mu_k &\geq \frac{\ln \mu_{k-1} + \ln \mu_{k+1}}{2} \\ \frac{-\ln \mu_{k-1} - \ln \mu_{k+1}}{2} &\geq -\ln \mu_k. \end{aligned}$$

Since this works for all $k \in [n-1]$, we have satisfied the hypotheses of Problem 12.31 for $a_i = \ln \mu_i$. So we can use $\mu_0 = 1$ to conclude that

$$\frac{-\ln \mu_k}{k} = \frac{-\ln \mu_k + \ln \mu_0}{k} = \frac{a_k - a_0}{k} \leq \frac{a_{k+1} - a_0}{k+1} = \frac{-\ln \mu_{k+1} + \ln \mu_0}{k+1} = \frac{-\ln \mu_{k+1}}{k+1}.$$

This is equivalent to

$$\begin{aligned} \ln\left[(\mu_{k+1})^{\frac{1}{k+1}}\right] &\leq \ln\left[(\mu_k)^{\frac{1}{k}}\right] \\ (\mu_{k+1})^{\frac{1}{k+1}} &\leq (\mu_k)^{\frac{1}{k}}, \end{aligned}$$

which is what we wanted. If equality holds, then it held in the application of Newton's inequality

$$\mu_k \geq (\mu_{k-1}\mu_{k+1})^{\frac{1}{2}}.$$

Since all of the r_i are positive, equality in Newton here requires $r_1 = r_2 = \cdots = r_n$. Conversely, if all of the r_i are equal, we leave it to the reader to work through the algebra that shows that equality holds in Maclaurin. □

Another inequality involving symmetric expressions is that of Muirhead. We have not included it because of the complexity of the proof. For a complete proof, see [2]. There are many further techniques that can be studied. The interested reader will find no deficit of resources from which such further results and problem-solving techniques can be learned for proving inequalities. For example, see the famous books, [7] and [8], as well as [10].

Appendices

Appendix A

Matrices

"When I look at the history of mathematics, I see a succession of illogical jumps, improbable coincidences, jokes of nature."

– *Freeman Dyson, Birds and Frogs*

"Mathematics is an experimental science, and definitions do not come first, but later on. They make themselves, when the nature of the subject has developed itself."

– *Oliver Heaviside, On operations in physical mathematics II*

We will outline here the definitions and basic properties of matrices that are used in the book.

Definition A.1. If m and n are positive integers, then, informally, an $m \times n$ **matrix** is an array or table of numbers with m rows and n columns. Formally, it is an indexing function whose indexing set is $[m] \times [n]$.

Definition A.2. The **matrix multiplication** of 2×2 matrices is defined by

$$\begin{pmatrix} a & b \\ c & d \end{pmatrix} \cdot \begin{pmatrix} p & q \\ r & s \end{pmatrix} = \begin{pmatrix} ap + br & aq + bs \\ cp + dr & cq + ds \end{pmatrix}.$$

Similarly, a 2×2 matrix times a column vector is computed as

$$\begin{pmatrix} a & b \\ c & d \end{pmatrix} \cdot \begin{pmatrix} p \\ q \end{pmatrix} = \begin{pmatrix} ap + bq \\ cp + dq \end{pmatrix}.$$

Matrix multiplication is associative, but not commutative. Positive integer **powers** of a 2×2 matrix are defined as repeated multiplication, just like the exponentiation of integers. The multiplication of larger matrices follows a similar definition. The multiplication of a matrix by a scalar (for us, a scalar is a real number but a more general definition involving field elements exists in abstract algebra), called **scalar multiplication**, is defined as

$$\alpha \cdot \begin{pmatrix} a & b \\ c & d \end{pmatrix} = \begin{pmatrix} \alpha a & \alpha b \\ \alpha c & \alpha d \end{pmatrix}.$$

We need matrix multiplication in the proof of the barycentric shoelace formula in Volume 4. The multiplication of a matrix with a column vector is useful as well for succinctly expressing a system of equations in the proof of the uniqueness of bivariate quadratic representations of conics in Volume 4. The powers of a matrix are used in a proof of Cassini's identity for the Fibonacci numbers in Volume 2.

Definition A.3. We call

$$I = \begin{pmatrix} 1 & 0 \\ 0 & 1 \end{pmatrix}$$

the 2×2 **identity matrix** because $IM = MI$ for every 2×2 matrix M, as the reader should verify. The identity matrix is like the number 1 in the multiplication of real numbers, since multiplying any matrix by I gives back that matrix.

Definition A.4. We will not define the **determinant** of an $n \times n$ matrix in general but the following formulas show how to compute the determinants of 2×2 and 3×3 matrices:

- The determinant of a 2×2 matrix is the real number

$$\det \begin{pmatrix} a & b \\ c & d \end{pmatrix} = ad - bc.$$

- The determinant of a 3×3 matrix is the real number

$$\det \begin{pmatrix} a & b & c \\ d & e & f \\ g & h & i \end{pmatrix} = aei + bfg + cdh - ceg - bdi - afh.$$

Theorem A.5. One way of remembering how to compute the 3×3 determinant, called the **Rule of Sarrus**, is shown by the following diagram:

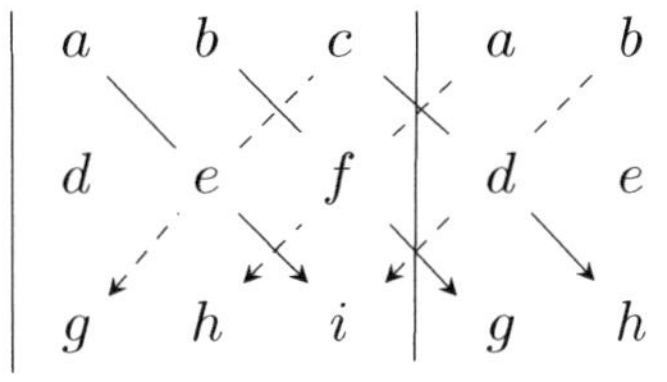

Theorem A.6. For 2×2 matrices M and N,

$$\det(M \cdot N) = \det(M) \cdot \det(N).$$

This is called the multiplicative property of the determinant.

Proof. We will use the matrices $M = \begin{pmatrix} a & b \\ c & d \end{pmatrix}$ and $N = \begin{pmatrix} p & q \\ r & s \end{pmatrix}$. Their product is

$$M \cdot N = \begin{pmatrix} ap + br & aq + bs \\ cp + dr & cq + ds \end{pmatrix}.$$

We find that

$$\begin{aligned} \det(M \cdot N) &= (ap + br)(cq + ds) - (aq + bs)(cp + dr) \\ &= adps + bcqr - adqr - bcps \\ &= (ad - bc)(ps - qr) \\ &= \det(M) \cdot \det(N). \end{aligned}$$

□

We will need the following properties that hold for the determinants of all square matrices, though we will need to apply them only to 2×2 and 3×3 matrices.

Theorem A.7. The facts below about determinants may be verified using the algebraic definitions (Definition A.4), albeit such direct proofs are tedious:

1. A scalar can be factored out of a single row or a single column of the determinant to outside the determinant without altering the value of the determinant. Equivalently, multiplying every element of one row or every element of one column by the same number results in multiplying the determinant by the same number.

2. Adding or subtracting (a scalar multiple) of a row from another row preserves the determinant. The same is true for the corresponding statement about columns.

3. Swapping two rows with each other or two columns with each other changes the sign of the determinant.

4. The determinant is preserved under taking the transpose of the matrix, meaning the element in each row i and column j is exchanged with the element in row j and column i. This operation is equivalent to reflecting the matrix across the diagonal that runs from the top-left to the bottom-right.

The determinant appears in a proof of the extended Euclidean algorithm in Volume 3, when studying lines from the perspective of coordinates in Volume 4, and in the derivation of various shoelace-style area formulas in Volume 4. Also, the Vandermonde matrix from Volume 2 and its determinant are relevant to the study of polynomials in Volume 1, and again in the theory of linear recurrences in Volume 2.

Appendix B

Solutions

"Yes, mathematics has two faces; it is the rigorous science of Euclid but it is also something else. Mathematics presented in the Euclidean way appears as a systematic, deductive science; but mathematics in the making appears as an experimental, inductive science. Both aspects are as old as the science of mathematics itself."

– *George Pólya, How to Solve It*

Solution 1.4. We can check that, for all four truth assignments of $t(p)$ and $t(q)$,

$$t(p \wedge q) = t(p) \cdot t(q),$$
$$t(p \vee q) = t(p) + t(q) - t(p) \cdot t(q).$$

Finally, we also find that $t(\neg p) = 1 - t(p)$.

Solution 1.14. We draw a master truth table including the premises and conclusions of all five arguments:

p	q	$\neg p$	$\neg q$	$p \to q$	$q \to p$	$q \to q$
0	0	1	1	1	1	1
0	1	1	0	1	0	1
1	0	0	1	0	1	1
1	1	0	0	1	1	1

We can see that, for each of the five fallacies, there is always a row in which the premises are true without the conclusion being true. The first three fallacies are discredited by row 2, and the last two fallacies are discredited by row 1. For the non-equivalence fallacy, we note that even though we cannot necessary deduce the truth of the converse from an implication, there are occasional examples in mathematics where knowing $P \to Q$ can help to prove $Q \to P$. We see an example of this when we use the Pythagorean theorem to prove its own converse in Volume 4.

Solution 1.23. We could derive the full truth tables for these two propositional variables, but it suffices to find just one assignment of truth values to p, q, r that causes the two expressions to evaluate differently. The very first possibility of all 0's works out because then, in informal notation,

$$p \to (q \to r) = 0 \to (0 \to 0) = 0 \to 1 = 1$$
$$(p \to q) \to r = (0 \to 0) \to 0 = 1 \to 0 = 0.$$

The two are in conflict, as expected.

Solution 1.34. Let T_1 be the valid argument H_1, C_1 and T_2 be the argument H_2, C_2. Since T_1 is stronger than T_2, the arguments H_2, H_1 and C_1, C_2 are both valid (if H_2 and H_1 are the same or if C_1 and C_2 are the same, this still works). Let us hypothetically construct a truth table with a column for each of H_2, H_1, C_1, C_2. Now we take three similar steps:

1. By the validity of H_2, H_1, whenever H_2 has a value of 1, a value of 1 is also assigned to H_1.
2. By the validity of H_1, C_1, whenever H_1 has a value of 1, a value of 1 is also assigned to C_1.

3. By the validity of C_1, C_2, whenever C_1 has a value of 1, a value of 1 is also assigned to C_2.

Putting it all together, whenever H_2 has a value of 1, a value of 1 is also assigned to C_2, proving that H_2, C_2 is valid. Thus, T_2 is a valid argument.

Solution 1.42. We will make assertions about a scenario where we have one blue marble and one red marble.

- It is true that "for every marble, the marble is blue or the marble is red." However, it is not true that "every marble is blue or every marble is red."
- It is not true that "there exists a marble that is both blue and red" but it is true that "there exists a blue marble and there exists a red marble."

Since the truth values do not match up, $\forall$ does not generally distribute over disjunction and $\exists$ does not generally distribute over conjunction.

Solution 1.46. It follows from the definitions of bounded variables with quantifiers that

$$\begin{aligned}\neg((\forall x, R(x)), P(x)) &= \neg(\forall x, (R(x) \to P(x))) = \neg(\forall x, (\neg R(x) \vee P(x))) \\ &= \exists x, (R(x) \wedge \neg P(x)) = (\exists x, R(x)), \neg P(x)\end{aligned}$$

and similarly,

$$\begin{aligned}\neg((\exists x, R(x)), P(x)) &= \neg(\exists x, (R(x) \wedge P(x))) = \forall x, (\neg R(x) \vee \neg P(x)) \\ &= \forall x(R(x) \to \neg P(x)) = (\forall x, R(x)), \neg P(x).\end{aligned}$$

Solution 2.11. We will use a sequence of set identities, as follows:

$$\begin{aligned}A^c &= \mathcal{U} \backslash A = \mathcal{U} \cap A^c \\ &= (A \cup B) \cap A^c = (A \cap A^c) \cup (B \cap A^c) \\ &= \varnothing \cup (B \cap A^c) = (B \cap A) \cup (B \cap A^c) \\ &= B \cap (A \cup A^c) = B \cap \mathcal{U} = B.\end{aligned}$$

Other proofs using alternative methods are possible, but this was the cleanest one that we could develop.

Solution 2.31. Suppose $f : X \to Y$ is a function, and that $g : Y \to X$ and $h : Y \to X$ are inverses of f. Then g is a left-inverse of f and h is a right-inverse of f. By Theorem 2.30, $g = h$.
Now, we will show that left-inverses and right-inverses are not necessarily unique.

- Let $f : \{0, 1\} \to \{0, 1, 2\}$ be defined by $0 \mapsto 0$ and $1 \mapsto 1$. Then there are the two distinct left-inverses $g : \{0, 1, 2\} \to \{0, 1\}$ and $h : \{0, 1, 2\} \to \{0, 1\}$, where both map $0 \mapsto 0$ and $1 \mapsto 1$, but g maps $2 \mapsto 0$ and h maps $2 \mapsto 1$.
- Let $f : \{0, 1\} \to \{0\}$ be defined by $0 \mapsto 0$ and $1 \mapsto 0$. Then there are the two distinct right-inverses $g : \{0\} \to \{0, 1\}$ and $h : \{0\} \to \{0, 1\}$, where g maps $0 \mapsto 0$ and h maps $0 \mapsto 1$.

Solution 2.35. Suppose $f : X \to Y$ and $g : Y \to Z$ are injective. Then for any two inputs $x, y \in X$,

$$(g \circ f)(x) = (g \circ f)(y) \implies g(f(x)) = g(f(y)) \implies f(x) = f(y) \implies x = y,$$

so $g \circ f$ is injective.
Separately, suppose $f : X \to Y$ and $g : Y \to Z$ are surjective. Then the range of f and the domain of g are both Y, which gives

$$(g \circ f)(X) = g(f(X)) = g(Y) = Z,$$

proving the surjectivity of $g \circ f$.
Bijectivity is also preserved under composition because a function is defined to be bijective if and only if it is both injective and surjective.

Solution 2.37. Suppose f is a cyclic function of order m. If $f(x) = f(y)$, then applying f by $(m-1)$ times to each side of the equation gives

$$x = f^m(x) = f^m(y) = y.$$

Thus, cyclicity implies injectivity.

Solution 2.38. Composing f with itself gives

$$f(f(x)) = \frac{\frac{x-1}{x} - 1}{\left(\frac{x-1}{x}\right)} = \frac{1}{1-x}.$$

Composing f over this again yields

$$f^3(x) = \frac{\frac{1}{1-x} - 1}{\left(\frac{1}{1-x}\right)} = x,$$

as desired. In the problem statement, we reverse-engineered the exclusion of both 0 and 1 from both the domain and codomain so that there is no division by 0 at any point of the compositions, and so that the codomain is a subset of the domain.

Solution 2.45. A function $f : X \to Y$ is an injection if and only if f has a left-inverse $g : Y \to X$. Next, g is a left-inverse of f if and only if f is a right-inverse of g. Finally, g has a right-inverse if and only if g is surjective, whose proof in Theorem 2.43 relies on the axiom of choice. As all of the steps were reversible, we are done.

Solution 2.46. We need g and h to disagree somewhere, but they need to agree on the range of f. Choosing $f(x) = x^2$ so that its range is $\mathbb{R}_{\geq 0}$, we find that we just need g, h to agree on the non-negative reals and disagree on some subset of the negative reals. So we choose $g(x) = x$ and $h(x) = |x|$. Incidentally, this also provides an example of functions f, g, h with domains and codomains $\mathbb{R}$ such that $f \circ g = f \circ h$ but $g \neq h$.

Solution 2.47. Injectivity without surjectivity is not difficult to achieve, say using an exponential function, such as $f(x) = 2^x$. For surjectivity without injectivity, we can choose a cubic function with two "bends," such as $g(x) = x(x-1)(x+1) = x^3 - x$.

Solution 2.48. Let f and g have the stated domains and codomains.

1. Suppose $f(x) = f(y)$ for some $x, y \in A$. Then applying g to both sides of the equation yields $g(f(x)) = g(f(y))$ or $(g \circ f)(x) = (g \circ f)(y)$. By the injectivity of $g \circ f$, we find that $x = y$, thereby proving the injectivity of f. For the counterexample to the injectivity of g, we first choose $g(x) = x^2$ as a classic non-injective function. Then we recognize that, in order for $g \circ f$ to be injective, it is necessary that there are not negatives in the range of f. Choosing $f(x) = 2^x$ closes the deal.

2. Since $g \circ f$ is surjective, we find that $(g \circ f)(A) = C$ or $g(f(A)) = C$. As a result,

$$f(A) \subseteq B \implies C = g(f(A)) \subseteq g(B) \subseteq C.$$

So $C \subseteq g(B)$ and $g(B) \subseteq C$, and antisymmetry gives $g(B) = C$, proving that g is surjective. For the counterexample to the surjectivity of f, we realize that it is necessary for f to map to only a subset of $\mathbb{R}$, yet g must be able to take this subset and map it to the entirety of $\mathbb{R}$. To that end, we choose $f(x) = 2^x$ so that its range is the positive reals, and we define $g(x) = \tan x$ on the domain of $\tan x$ and $g(x) = 0$ (this is an arbitrary value) wherever $\tan x$ has a vertical asymptote. Since $\tan x$ takes, for example, $\left(\dfrac{\pi}{2}, \dfrac{3\pi}{2}\right)$ and maps it to the entirety of $\mathbb{R}$, this does the job.

Solution 2.49. An initial clue is that neither f nor g can be a bijection, for otherwise the bijection f^{-1} or g^{-1} could be right-composed or left-composed with $g \circ f$ to get that g or f is bijective, respectively; this is impossible because then we could compose $f \circ g$ in either case to get that $f \circ g$ is a bijection, thwarting our efforts. Moreover, by Problem 2.48, if $g \circ f$ is bijective, its injectivity implies the injectivity of f, and its surjectivity implies the surjectivity of g. Putting it all together, we try $f(x) = 2^x$ as it is injective but not surjective. Choosing $g(x) = \log_2 x$ would give the bijection $(g \circ f)(x) = x$, but we still need to define g on the non-positive reals and ensure that $f \circ g$ is not bijective. To simultaneously satisfy these goals, we define $g(x) = 0$ for $x \leq 0$. This does the trick.

Solution 2.51. We will develop bijections, explicitly or implicitly, with domains and codomains $[0, 1] \to (0, 1)$, $(0, 1) \to \mathbb{R}$, $\mathbb{R} \to (0, \infty)$, and $(0, \infty) \to [0, \infty)$.

1. Making use of Schröder-Bernstein (Theorem 2.50), we will produce an injection from $[0, 1]$ to $(0, 1)$ and vice versa. These are $x \mapsto \dfrac{x}{2} + \dfrac{1}{4}$ and simply $x \mapsto x$.
2. A bijection from $(0, 1)$ to $\mathbb{R}$ is given by $x \mapsto \tan\left[\pi\left(x - \dfrac{1}{2}\right)\right]$.
3. A bijection from $\mathbb{R}$ to $(0, \infty)$ is given by $x \mapsto 2^x$.
4. Again, we will make use of Schröder-Bernstein by producing an injection from $(0, \infty)$ to $[0, \infty)$ and vice versa. These maps are $x \mapsto x$ and $x \mapsto x + 1$.

By transitivity of the "same cardinality" property (which may be proven by taking compositions of bijections), we are done.

Solution 2.56. We label each sticker so that it has a unique identifier (while retaining its colour). Although not every rearrangement of the stickers is achievable by legal moves, every achievable rearrangement can be considered to be a bijection from the possible locations on the six faces to the set of labelled stickers. If five of the faces are solved, meaning there is a bijection from yellow inputs to yellow outputs, and similarly for four more colours, then that leaves only 9 stickers of the same colour for the final face. Since every achievable rearrangement is a bijection as described, the final face should be automatically solved.

Solution 2.60. Let X be a non-empty subset of $\mathbb{Z}$ with a lower bound b. Since $x \geq b$ for all $x \in X$, it holds that $x - b + 1 \geq 1$ and so all elements of $X - b + 1 = \{x - b + 1 : x \in X\}$ are in $\mathbb{Z}_+$. By the assumed weak version of the well-ordering principle, $X - b + 1$ has a least element m. So $x - b + 1 \geq m$ for all $x \in X$. Thus,

$$\forall x \in X, x \geq m + b - 1.$$

All we have to do is prove that $m + b - 1 \in X$. Indeed, since $m \in X - b + 1$, there exists an $x_0 \in X$ such that $m = x_0 - b + 1$. Therefore,

$$m + b - 1 = x_0 \in X.$$

Solution 2.63. The base case $n = 1$ is easy to verify. Afterwards, suppose

$$\frac{1}{1^2} + \frac{1}{2^2} + \frac{1}{3^2} + \cdots + \frac{1}{n^2} \leq 2 - \frac{1}{n}$$

holds for some positive integer n. Adding $\dfrac{1}{(n+1)^2}$ to both sides yields

$$\frac{1}{1^2} + \frac{1}{2^2} + \frac{1}{3^2} + \cdots + \frac{1}{n^2} + \frac{1}{(n+1)^2} \leq 2 - \frac{1}{n} + \frac{1}{(n+1)^2}.$$

It suffices to prove that

$$2 - \frac{1}{n} + \frac{1}{(n+1)^2} \leq 2 - \frac{1}{n+1}.$$

Working backwards by clearing the denominators and expanding reduces it to $0 \leq 1$, which is true. (It is actually known that the infinite sum is exactly $\frac{\pi^2}{6}$.)

Solution 2.64. The base case for $n = 1$ may be directly verified. Now, suppose there exists a positive integer n such that

$$\sqrt{n} \leq \frac{1}{\sqrt{1}} + \frac{1}{\sqrt{2}} + \frac{1}{\sqrt{3}} + \cdots + \frac{1}{\sqrt{n}} \leq 2\sqrt{n}.$$

Adding $\frac{1}{\sqrt{n+1}}$ to all three components yields

$$\sqrt{n} + \frac{1}{\sqrt{n+1}} \leq \frac{1}{\sqrt{1}} + \frac{1}{\sqrt{2}} + \frac{1}{\sqrt{3}} + \cdots + \frac{1}{\sqrt{n}} + \frac{1}{\sqrt{n+1}} \leq 2\sqrt{n} + \frac{1}{\sqrt{n+1}}.$$

By transitivity of inequalities, it suffices to weaken the inequalities by proving that

$$\sqrt{n+1} \leq \sqrt{n} + \frac{1}{\sqrt{n+1}},$$
$$2\sqrt{n} + \frac{1}{\sqrt{n+1}} \leq 2\sqrt{n+1}.$$

Working backwards by clearing denominators, cancelling or collecting like terms, and squaring both sides as needed, we find that the former is equivalent to $0 \leq n$ and the latter is equivalent to $0 \leq 1$, both of which are true. Thus, the induction is complete. Strangely, we had to weaken our result in the inductive step by widening the gaps between the parts of the inequality in order to have the effect of bringing the induction dominoes closer together and cause a domino chain reaction.

Solution 2.65. By computing the first few cases, we suspect that the formula is

$$\frac{1}{2!} + \frac{2}{3!} + \frac{3}{4!} + \cdots + \frac{n}{(n+1)!} = 1 - \frac{1}{(n+1)!}.$$

The base case $n = 1$ is easy enough to verify. Then, assuming that the formula holds for some positive integer n, we add $\frac{n+1}{(n+2)!}$ to both sides to get

$$\frac{1}{2!} + \frac{2}{3!} + \frac{3}{4!} + \cdots + \frac{n}{(n+1)!} + \frac{n+1}{(n+2)!} = 1 - \frac{1}{(n+1)!} + \frac{n+1}{(n+2)!} = 1 - \frac{(n+2)-(n+1)}{(n+2)!} = 1 - \frac{1}{(n+2)!},$$

as desired. This completes the induction.

Solution 3.5. The following are statements that apply to all elements a, b, c of the underlying set:

- Commutativity of $\circ$: $\circ(a, b) = \circ(b, a)$
- Associativity of $\circ$: $\circ(a, \circ(b, c)) = \circ(\circ(a, b), c)$
- Left-distributivity of $\star$ over $\circ$: $\star(c, \circ(a, b)) = \circ(\star(c, a), \star(c, b))$
- Right-distributivity of $\star$ over $\circ$: $\star(\circ(a, b), c) = \circ(\star(a, c), \star(b, c))$

This shows how prefix function notation is unwieldy in the case of binary operations and is discarded as such.

Solution 3.28. We find that

$$-(-a) = (-1)((-1)a) = ((-1)(-1))a = (1 \cdot 1)a = 1 \cdot a = a.$$

We have used Corollary 3.27 to compute $(-1)(-1) = 1 \cdot 1 = 1$. Alternatively, $-a$ is defined as the number such that $a + (-a) = 0$. But then a is the number such that $(-a) + a = 0$, so the negative of $-a$ is a itself.

Solution 3.32. Suppose $f : \mathbb{R} \to \mathbb{R}$ is a function that is both even and odd. Then for all real x,

$$-f(x) = f(-x) = f(x) \implies 0 = f(x) + f(x) = 2f(x).$$

Thus, the only possibility is that $f(x) = 0$ for all real x, Indeed, this function is both even and odd.

Solution 3.38. We know that $(-1)^{-1} = -1$ because

$$(-1)(-1) = 1 \cdot 1 = 1,$$

which leads to

$$\frac{-1}{-1} = (-1) \cdot (-1)^{-1} = (-1)(-1) = 1.$$

Solution 3.40. Since a^{-1} is defined as the number such that $a \cdot a^{-1} = 1$, we find that a is the number such that $a^{-1} \cdot a = 1$. So the inverse of a^{-1} is a.

Solution 3.44. The identities are derived from previous results as follows:

- $\frac{-a}{-b} = \frac{(-1)a}{(-1)b} = \frac{-1}{-1} \cdot \frac{a}{b} = \frac{a}{b}$
- $\frac{-a}{b} = \frac{(-1)a}{(1)b} = \frac{-1}{1} \cdot \frac{a}{b} = (-1) \cdot \frac{a}{b} = -\frac{a}{b}$ and the second equality follows similarly

Solution 3.47. We have to verify reflexivity, symmetry and transitivity. Before we get into the proof, we will prove a lemma, which states that if $(a, b) \sim (c, 0)$ then $b = 0$, and if $(a, b) \sim (0, d)$ then $a = 0$. If $(a, b) \sim (c, 0)$ then $0 = a \cdot 0 = bc$. Since $(0, 0)$ is not a part of the set, $c \neq 0$ and so $b = 0$. For the second assertion, if $(a, b) \sim (0, d)$ then $ad = b \cdot 0 = 0$. Again since $(0, 0)$ is not a part of the set, $d \neq 0$ and so $a = 0$.

1. Reflexivity: $(a, b) \sim (a, b)$ is true because it is means $ab = ba$, which holds by commutativity of multiplication
2. Symmetry: Suppose $(a, b) \sim (c, d)$. Then $ad = bc$. This is equivalent to $cb = da$, which implies $(c, d) \sim (a, b)$.
3. Transitivity: Suppose $(a, b) \sim (c, d)$ and $(c, d) \sim (e, f)$. Then $ad = bc$ and $cf = de$. We want to show that $af = be$, as that would give us $(a, b) \sim (e, f)$. We look at three cases:
   - Suppose $c \neq 0$ and $d \neq 0$. Multiplying $ad = bc$ and $cf = de$ yields $adcf = bcde$. We can divide both sides by cd to get $af = be$.
   - If $c = 0$ then $a = e = 0$ by the lemma, and so $af = 0 = be$.
   - If $d = 0$ then $b = f = 0$ by the lemma, and so $af = 0 = be$ again.

Therefore, the ratio binary binary relation is an equivalence relation.

Solution 3.48. Neither function is the identity function, so neither function is of order 1. So it suffices to show that $f \circ f$ and $g \circ g$ are both the identity function. Indeed,

$$f(f(x)) = c - (c - x) = c - c + x = x,$$
$$g(g(x)) = \frac{c}{\left(\frac{c}{x}\right)} = c \cdot \frac{x}{c} = x.$$

Solution 3.57. Our key tool will be the difference of squares factorization in the form stated in the problem. For the first one,

$$\frac{1}{\sqrt{2} + \sqrt{3}} = \frac{\sqrt{2} - \sqrt{3}}{(\sqrt{2} + \sqrt{3})(\sqrt{2} - \sqrt{3})} = \frac{\sqrt{2} - \sqrt{3}}{2 - 3} = \sqrt{3} - \sqrt{2}.$$

For the second fraction, we will apply the difference of squares twice:

$$\begin{aligned}\frac{1}{\sqrt{2}-\sqrt{3}+\sqrt{5}} &= \frac{\sqrt{2}+\sqrt{5}+\sqrt{3}}{(\sqrt{2}+\sqrt{5}-\sqrt{3})(\sqrt{2}+\sqrt{5}+\sqrt{3})}\\ &= \frac{\sqrt{2}+\sqrt{3}+\sqrt{5}}{2(2+\sqrt{10})}\\ &= \frac{(\sqrt{2}+\sqrt{3}+\sqrt{5})(2-\sqrt{10})}{2(2+\sqrt{10})(2-\sqrt{10})}\\ &= \frac{(\sqrt{2}+\sqrt{3}+\sqrt{5})(2-\sqrt{10})}{2(4-10)}\\ &= \frac{(\sqrt{2}+\sqrt{3}+\sqrt{5})(\sqrt{10}-2)}{12}.\end{aligned}$$

Solution 4.6. Again, the answer is not $40-10=30$. The list of pages is $(11,12,13,\dots,39)$. We subtract 10 from each entry to turn it into the list $(1,2,3,\dots,29)$. So there are 29 pages from page 10 to page 40 exclusive.

Solution 4.7. We begin by finding the first and last multiples on 3 in the given interval. By using long division, we find that $110 = 3\cdot 36+2$ and $1000 = 3\cdot 333+1$. So the list of relevant multiples of 3 is

$$(3\cdot 37, 3\cdot 38, 3\cdot 39, \dots, 3\cdot 333).$$

We divide every entry by 3 to produce the list $(37,38,39,\dots,333)$, and then we subtract 36 from each of these entries to get $(1,2,3,\dots,297)$. Thus, the answer is 297.

Solution 4.8. The two solutions apply injections to lists, as usual.

1. We subtract a from each term and then divide each resulting term by d to get the list $(n,n+1,\dots,m)$. Then we subtract $n-1$ from each term to get $(1,2,\dots,m-n+1)$.

2. We divide each term by br^{n-1} to get the list

$$(r^1, r^2, \dots, r^{m-n+1}).$$

We take the absolute value of each term, and then we apply the $\log_{|r|} x$ function to each term to get the list

$$(1,2,\dots,m-n+1).$$

In both cases, the number of indices is $m-n+1$. There is no need to memorize this result, as knowledge of the proof will allow the reader to easily replicate the process.

Solution 4.9. Conveniently, $(-10)^3 = -1000$ and $10^3 = 1000$, and the next cubes are too large in magnitude in either direction. So the list of relevant cubes of integers, ordered from least to greatest, is

$$((-10)^3, (-9)^3, (-8)^3, \dots, 8^3, 9^3, 10^3).$$

By taking the cube root of each entry, we get the list

$$(-10,-9,-8,\dots,8,9,10),$$

which has $10+1+9=10=21$ entries in it. We did the final computation by noticing that there are 10 negative entries, one 0 entry, and 10 positive entries.

Solution 4.12. The idea is to make the counter what we are subtracting from t at each step. That is,

$$\begin{aligned}\sum_{i=s}^{t} a_j &= a_s + a_{s+1} + \cdots + a_{t-1} + a_t = a_t + a_{t-1} + \cdots + a_{s+1} + a_s \\ &= a_t + a_{t-1} + \cdots + a_{t-(t-s-1)} + a_{t-(t-s)} = \sum_{j=0}^{t-s} a_{t-j}.\end{aligned}$$

This is known as a *reflection*, since we are reflecting the indices from right to left.

Solution 4.20. As in the proof of Theorem 4.18, the terms can be placed in a matrix:

$$\begin{bmatrix} a_{12} & a_{13} & \cdots & a_{1n} \\ & a_{23} & \cdots & a_{2n} \\ & & \ddots & \vdots \\ & & & a_{(n-1)n} \\ & & & \end{bmatrix}$$

The left side of the identity finds the sum of the matrix by adding the sums of rows, and the right side is the sums of columns.

Solution 5.4. We know that $-2 \neq 2$, but squaring both sides gives $4 \neq 4$, which is untrue since we know that $4 = 4$. Technically, we applied the function $f(t) = t^2$ to both sides.

Solution 5.8. We will use the smashing technique. Multiplying the three equations together yields

$$pqr = \left(x + \frac{1}{y}\right)\left(y + \frac{1}{z}\right)\left(z + \frac{1}{x}\right) = xyz + x + y + z + \frac{1}{x} + \frac{1}{y} + \frac{1}{z} + \frac{1}{xyz}.$$

Adding the three equations together yields

$$p + q + r = x + y + z + \frac{1}{x} + \frac{1}{y} + \frac{1}{z}.$$

Therefore, we can subtract the added equation from the multiplied equation to get

$$pqr - (p + q + r) = xyz + \frac{1}{xyz}.$$

Solution 5.9. Let a be a real number that is its own additive inverse. Then $a + a = 0$. By the distributive property,

$$2a = a \cdot (1 + 1) = 0.$$

Dividing both sides by 2, we get that $a = 0$, which indeed works. So 0 is the only solution.
Let b be a real number that is its own multiplicative inverse. Then $b^2 = b \cdot b = 1$. We will use the difference of squares factorization, which, using the distributive property, says

$$(x - y)(x + y) = x \cdot x + x \cdot y - y \cdot x - y \cdot y = x^2 - y^2.$$

Then $b^2 = 1$ can be rewritten as

$$0 = b^2 - 1^2 = (b - 1)(b + 1).$$

Thus either $b - 1 = 0$ or $b + 1 = 0$, by Theorem 3.36. This means the potential solutions are 1 and -1. They are indeed their own inverses, as we can verify by computation: $(-1) \cdot (-1) = 1 = 1 \cdot 1$.

Solution 5.10. The errors occurs when we divides both sides of

$$(a - b)(a + b) = (a - b)b$$

by $a - b$. Since we started with the assumption that $a = b$, this means dividing by $a - b$ is the same as dividing by 0, which has no definition.

Solution 5.17. The idea is to change one variable at a time. By the compatibility rules (Theorem 5.16), we find that:

$$\begin{aligned} a + c &> b + c > b + d, \\ a + c &\geq b + c > b + d, \\ a + c &\geq b + c \geq b + d, \\ ac &> bc > bd. \end{aligned}$$

For further results, just go step by step and use one compatibility rule at a time.

Solution 5.21. We choose the sequence that takes the reciprocals of the integers: $\frac{1}{1}, \frac{1}{2}, \frac{1}{3}, \frac{1}{4}, \ldots$. Suppose $\epsilon > 0$ is fixed. We want a positive integer N such that

$$\frac{1}{N}, \frac{1}{N+1}, \frac{1}{N+2}, \frac{1}{N+3}, \ldots$$

are all less than ϵ. An equivalent condition is

$$\frac{1}{\epsilon} < N, N+1, N+2, \ldots.$$

By the Archimedean property, there exists an integer N such that $\frac{1}{\epsilon} < N$. It follows from transitivity that all subsequent integers are also greater than $\frac{1}{\epsilon}$.

Solution 5.23. Any open interval such as $(0, 1)$ works. An upper bound is 1 and a lower bound is 0, but neither lives within the set. Since the upper bound 1 is the supremum and the lower bound 0 is the infimum, there is no maximum or minimum.

Solution 5.30. By multiplying together all of the inequalities

$$a_1 \geq b_1, a_2 \geq b_2, \ldots, a_n \geq b_n,$$

we find that

$$a_1 a_2 \cdots a_n \geq b_1 b_2 \cdots b_n$$

is true.

For the equality criterion, an idea is to use the fact that $\log_{10} x$ is a strictly increasing function with its domain being the positive reals and its range being the reals (in fact, any increasing function $\log_b x$, for a fixed base $b > 1$, will work). This will allow us to reduce this problem to its additive variant, which is one of the parts of Lemma 5.29. If

$$a_1 = b_1, a_2 = b_2, \ldots, a_n = b_n,$$

then we can multiply them together to find that equality holds in our inequality. Conversely, suppose

$$a_1 a_2 \cdots a_n = b_1 b_2 \cdots b_n.$$

By taking the logarithm and using the product rule, this can be rewritten as

$$\log_{10} a_1 + \cdots + \log_{10} a_n = \log_{10} b_1 + \cdots + \log_{10} b_n.$$

Since a logarithm to a fixed base $b > 1$ is a strictly increasing function,

$$a_k \geq b_k \implies \log_{10} a_k \geq \log_{10} b_k$$

for each $k \in [n]$. By the additive variant, $\log_{10} a_k = \log_{10} b_k$ for each $k \in [n]$, and so $a_k = b_k$ as well.

Solution 5.35. The inequality can be manipulated using reversible steps as follows:

$$\frac{x^2+y^2}{2} \geq xy \implies x^2+y^2 \geq 2xy \implies x^2-2xy+y^2 \geq 0 \implies (x-y)^2 \geq 0,$$

which is true by the trivial inequality. Equality holds if and only if $x = y$.

Solution 5.40. We need to prove connexity, antisymmetry, and transitivity of the $\leq$ relation defined.

1. Connexity: We wish to show that $a \leq b$ or $b \leq a$. Logically, it is equivalent to show that if $a \leq b$ is false, then $b \leq a$. Supposing $a \not\leq b$, the trichotomy law implies that the only remaining option is $b < a$. This implies $b \leq a$.

2. Antisymmetry: Suppose $a \leq b$ and $b \leq a$. By definition, this means that $a < b$ or $a = b$, and $b < a$ or $b = a$. Suppose, for contradiction, that $a \neq b$. Then we are forced into saying that $a < b$ and $b < a$ at the same time. This is impossible by the asymmetry of a strict total order. Thus, $a = b$.

3. Transitivity: Suppose $a \leq b$ and $b \leq c$. By definition, this means that $a < b$ or $a = b$, and $b < c$ or $b = c$. This produces four cases:

    - If $a < b$ and $b < c$, then $a < c$.
    - If $a < b$ and $b = c$, then $a < c$.
    - If $a = b$ and $b < c$, then $a < c$.
    - If $a = b$ and $b = c$, then $a = c$.

    In all four conclusions, it is true that $a \leq c$.

Solution 5.44. Since we are using shortlex order, we can group the strings according to their length:

$$\begin{aligned}&0, 1,\\ &00, 01, 10, 11,\\ &000, 001, 010, 011, 100, 101, 110, 111.\end{aligned}$$

Note that, among strings with the same number of symbols, we have followed lexicographical order. We know that we have found all the strings because a combinatorial multiplication argument implies that there should be 2^n strings of each length n, which matches our results for $n = 1, 2, 3$.

Solution 6.6. It is possible to do this by induction, but we will simply square both sides as before. The reason is that we want to display a nice technique for deriving the equality condition when there are numerous terms. Squaring both sides is reversible since both sides are non-negative, and doing it yields

$$|x_1| + |x_2| + \cdots + |x_n| \geq |x_1 + x_2 + \cdots + x_n|$$

$$\sum_{k=1}^{n} |x_k|^2 + \sum_{1 \leq i < j \leq n} 2|x_i| \cdot |x_j| \geq \sum_{k=1}^{n} x_k^2 + \sum_{1 \leq i < j \leq n} 2x_i x_j$$

$$\sum_{1 \leq i < j \leq n} (|x_i x_j| - x_i x_j) \geq 0.$$

Now, we will use Lemma 5.29. Notice that each term $(|x_i x_j| - x_i x_j)$ is non-negative. In order for the sum of all such terms to equal 0, each term must be 0. The reason is that if a single such term is positive instead of 0, the whole sum will be positive instead of 0. So we must have $|x_i x_j| = x_i x_j$ for each pair of indices $1 \leq i < j \leq n$. This means x_i and x_j are both non-negative or both non-positive. Applying this rule to the $n-1$ pairs of numbers $(x_1, x_2), (x_2, x_3), \ldots, (x_{n-1}, x_n)$, we find that all of the x_k are non-negative or all of them are non-positive.

Solution 6.10. These identities can be proven by casework on the parity of n. We leave the computations for even $n = 2m$ and odd $n = 2m + 1$ to the reader.

Solution 6.15. Let $m = \lfloor x \rfloor$ and $n = \lceil -x \rceil$. We know that

$$\begin{aligned} x - 1 < m &\leq x, \\ -x \leq n < -x + 1 \iff x - 1 < -n &\leq x. \end{aligned}$$

This means m and $-n$ are both integers in the interval $(x - 1, x]$, in which we know there exists exactly one integer. Thus, $m = -n$.

Solution 6.17. Let $p = \left\lfloor \frac{m}{n} \right\rfloor$. This equation is equivalent to

$$p \leq \frac{m}{n} < p + 1 \iff pn \leq m < (p + 1)n \iff pn \leq m \leq (p + 1)n - 1,$$

where we have used the fact that $n > 0$ to clear the denominator. Similarly, let $q = \left\lceil \frac{m}{n} \right\rceil$. Then this equation is equivalent to

$$q - 1 < \frac{m}{n} \leq q \iff (q - 1)n < m \leq qn \iff (q - 1)n + 1 \leq m \leq qn.$$

Finally, we add or subtract $n - 1$ from all components to get:

$$\begin{aligned} pn \leq m \leq (p + 1)n - 1 &\iff (p - 1)n + 1 \leq m - n + 1 \leq pm, \\ (q - 1)n + 1 \leq m \leq qn &\iff qn \leq m + n - 1 \leq (q + 1)n - 1, \end{aligned}$$

which is equivalent to what we wanted to prove. We are done because all of our work was reversible.

Solution 7.5. This is an arithmetic series with initial term 1 and last term $2n - 1$, with a total of n terms. The formula yields

$$\left(\frac{1 + (2n - 1)}{2} \right) \cdot n = n^2.$$

Remarkably, the sum of the first n odd positive integers is the nth square.

Solution 7.8. By expanding, we notice that the expression

$$\underbrace{(\ldots((pr + p)r + p)r \ldots + p)r + p}_{\text{number of multiplications by } r \text{ is } n-1}$$

is equal to the geometric series

$$pr^{n-1} + pr^{n-2} + \cdots + pr^2 + pr + p = \frac{p(r^n - 1)}{r - 1}.$$

This pattern may be formally proven by induction, if desired. The original expression is the natural starting point of the so-called future value formula for annuities in finance, except r is replaced by $r = s + 1$ so that the formula can be written as

$$p \cdot \frac{(s + 1)^n - 1}{s}.$$

Solution 7.14. Using the key fact $(n + 1)! = (n + 1) \cdot n!$, we find that

$$\begin{aligned} 1 \cdot 1! + 2 \cdot 2! + 3 \cdot 3! + \cdots + n \cdot n! &= (2 - 1) \cdot 1! + (3 - 1) \cdot 2! + (4 - 1) \cdot 3! + \cdots + [(n + 1) - 1] \cdot n! \\ &= (2! - 1!) + (3! - 2!) + (4! - 3!) + \cdots + [(n + 1)! - n!] \\ &= (n + 1)! - 1, \end{aligned}$$

where we had telescoping in the final step.

Solution 7.17. Each term is of the form

$$\frac{1}{k(k+1)(k+2)} = \frac{1}{2}\left(\frac{1}{k} - \frac{2}{k+1} + \frac{1}{k+2}\right),$$

which we can find through partial fraction decomposition. There is a way to telescope the series by stacking the terms on top of each other and noticing that terms on diagonals cancel out because

$$\frac{1}{k} - \frac{2}{k} + \frac{1}{k} = 0.$$

Drawing it out and filling in the details are left as an exercise to the reader. We will instead show a cleverer method that uses an incomplete partial fraction decomposition:

$$\frac{1}{k(k+1)(k+2)} = \frac{1}{2}\left(\frac{1}{k(k+1)} - \frac{1}{(k+1)(k+2)}\right) = f(k) - f(k+1),$$

where $f(k) = \frac{1}{2} \cdot \frac{1}{k(k+1)}$. Then the nth partial sum is

$$\begin{aligned}\sum_{k=1}^{n} \frac{1}{k(k+1)(k+2)} &= \sum_{k=1}^{n} (f(k) - f(k+1)) = f(1) - f(n+1) \\ &= \frac{1}{2} \cdot \left[\frac{1}{1 \cdot 2} - \frac{1}{(n+1)(n+2)}\right] = \frac{1}{4} - \frac{1}{2(n+1)(n+2)},\end{aligned}$$

which goes to $\frac{1}{4}$ as n goes to infinity because the second term goes to 0.

Solution 7.19. It is easily seen that the base case $n = 1$ holds. Now, suppose that

$$1^3 + 2^3 + \cdots + n^3 = \left[\frac{n(n+1)}{2}\right]^2$$

for some integer $n \geq 1$. Then adding $(n+1)^3$ to both sides yields

$$\begin{aligned}1^3 + 2^3 + \cdots + n^3 + (n+1)^3 &= \frac{n^2(n+1)^2}{4} + (n+1)^3 = \frac{n^2(n+1)^2 + 4(n+1)^3}{4} \\ &= \frac{(n+1)^2[n^2 + 4(n+1)]}{4} = \frac{(n+1)^2(n+2)^2}{4} \\ &= \left[\frac{(n+1)(n+2)}{2}\right]^2,\end{aligned}$$

as desired.

Solution 7.21. Using the difference of cubes and sum of cubes formulas

$$\begin{aligned}k^3 - 1 &= (k-1)(k^2 + k + 1), \\ k^3 + 1 &= (k+1)(k^2 - k + 1),\end{aligned}$$

the partial product up to any specific index $n \geq 2$ is equal to

$$\prod_{k=2}^{n} \frac{k^3 - 1}{k^3 + 1} = \prod_{k=2}^{n} \frac{(k-1)(k^2 + k + 1)}{(k+1)(k^2 - k + 1)} = \left(\prod_{k=2}^{n} \frac{k-1}{k+1}\right)\left(\prod_{k=2}^{n} \frac{k^2 + k + 1}{k^2 - k + 1}\right).$$

Now we can evaluate each product separately and then bring them back together. The first product easily telescopes as

$$\prod_{k=2}^{n} \frac{k-1}{k+1} = \frac{1}{3} \cdot \frac{2}{4} \cdot \frac{3}{5} \cdots \frac{n-1}{n+1} = \frac{1 \cdot 2}{n(n+1)}.$$

The second product also telescopes, but due to the unexpected identity

$$k^2 + k + 1 = (k+1)^2 - (k+1) + 1.$$

Letting $f(k) = k^2 - k + 1$, the second product is

$$\prod_{k=2}^{n} \frac{k^2+k+1}{k^2-k+1} = \prod_{k=2}^{n} \frac{f(k+1)}{f(k)} = \frac{f(3)}{f(2)} \cdot \frac{f(4)}{f(3)} \cdots \frac{f(n+1)}{f(n)} = \frac{f(n+1)}{f(2)}.$$

Therefore, the infinite product is

$$\prod_{k=2}^{\infty} \frac{k^3-1}{k^3+1} = \lim_{n\to\infty} \prod_{k=2}^{n} \frac{k^3-1}{k^3+1} = \lim_{n\to\infty} \left[\frac{1\cdot 2}{n(n+1)} \cdot \frac{f(n+1)}{f(2)}\right] = \lim_{n\to\infty} \frac{2((n+1)^2-(n+1)+1)}{n(n+1)(2^2-2+1)}$$
$$= \frac{2}{3} \cdot \lim_{n\to\infty} \frac{n^2+n+1}{n^2+n} = \frac{2}{3} \cdot \lim_{n\to\infty} \left(1 + \frac{1}{n^2+n}\right).$$

As n goes to infinity, $\frac{1}{n^2+n}$ goes to 0, so the answer is $\frac{2}{3}$.

Solution 8.9. We will use the fact that $\tan : \left(-\frac{\pi}{2}, \frac{\pi}{2}\right) \to \mathbb{R}$ is a bijection. It suffices to find a bijection between every pair of open intervals, since the composition of bijections is a bijection. Let (a,b) and (c,d) be open intervals. The idea is to shift (a,b) over to $(0, b-a)$, scale that to $(0, d-c)$ and then shift that over to (c,d). Altogether, this produces the function

$$f(x) = (x-a) \cdot \frac{d-c}{b-a} + c.$$

This is is a linear function, so it is bijective on $\mathbb{R}$. We can verify that

$$a < x < b \iff c < f(x) < d,$$

so restricting f to the domain (a,b) makes the range (c,d). This completes the proof because restrictions of bijections are also bijections onto the new range.

Solution 8.18. By the addition identities and double angle identities for sine and cosine,

$$\begin{aligned}
\sin(3\theta) &= \sin(\theta + 2\theta) \\
&= \sin\theta\cos(2\theta) + \cos\theta\sin(2\theta) \\
&= \sin\theta(1 - 2\sin^2\theta) + \cos\theta(2\sin\theta\cos\theta) \\
&= \sin\theta - 2\sin^3\theta + 2\sin\theta\cos^2\theta \\
&= \sin\theta - 2\sin^3\theta + 2\sin\theta(1-\sin^2\theta) \\
&= \sin\theta - 2\sin^3\theta + 2\sin\theta - 2\sin^3\theta \\
&= 3\sin\theta - 4\sin^3\theta
\end{aligned}$$

and similarly,

$$\begin{aligned}
\cos(\theta) &= \cos(\theta + 2\theta) \\
&= \cos\theta\cos(2\theta) - \sin\theta\sin(2\theta) \\
&= \cos\theta(2\cos^2\theta - 1) - \sin\theta(2\sin\theta\cos\theta) \\
&= 2\cos^3\theta - \cos\theta - 2\sin^2\theta\cos\theta \\
&= 2\cos^3\theta - \cos\theta - 2(1-\cos^2\theta)\cos\theta \\
&= 2\cos^3\theta - \cos\theta - 2\cos\theta + 2\cos^3\theta \\
&= 4\cos^3\theta - 3\cos\theta.
\end{aligned}$$

As a result, using $\sec^2\theta = 1+\tan^2\theta$,

$$\begin{aligned}\tan(3\theta) &= \frac{\sin(3\theta)}{\cos(3\theta)} = \frac{3\sin\theta - 4\sin^3\theta}{4\cos^3\theta - 3\cos\theta} = \frac{\frac{3\sin\theta}{\cos^3\theta} - 4\frac{\sin^3\theta}{\cos^3\theta}}{4 - \frac{3}{\cos^2\theta}} \\ &= \frac{3\tan\theta\sec^2\theta - 4\tan^3\theta}{4 - 3\sec^2\theta} = \frac{3\tan\theta(1+\tan^2\theta) - 4\tan^3\theta}{4 - 3(1+\tan^2\theta)} = \frac{3\tan\theta - \tan^3\theta}{1 - 3\tan^2\theta}.\end{aligned}$$

Solution 9.9. The first few powers of i are: $i^1 = i, i^2 = -1, i^3 = -i, i^4 = 1$, and it can be shown by induction that this pattern cycles. We can also perform an induction on non-positive integer exponents that descend towards $-\infty$ to prove that the periodic pattern extends in the other direction: $i^0 = 1, i^{-1} = -i, i^{-2} = -1, i^{-3} = i$, and so on.

Solution 9.15. By the formula for a complex geometric series (Theorem 9.14), we can evaluate this series as

$$\begin{aligned}z^\alpha + z^{\alpha-\beta} + z^{\alpha-2\beta} + \cdots + z^\gamma &= z^\alpha \cdot \left(1 + \frac{1}{z^\beta} + \frac{1}{z^{2\beta}} + \cdots + \frac{1}{z^{q\beta}}\right) \\ &= z^\alpha \cdot \frac{1 - \left(\frac{1}{z^\beta}\right)^{q+1}}{1 - \frac{1}{z^\beta}} = z^\alpha \cdot \frac{z^{q\beta}\cdot z^\beta - 1}{z^{q\beta}(z^\beta - 1)} \\ &= z^\alpha \cdot \frac{z^{\alpha-\gamma}\cdot z^\beta - 1}{z^{\alpha-\gamma}(z^\beta - 1)} = \frac{z^{\alpha+\beta} - z^\gamma}{z^\beta - 1}.\end{aligned}$$

Solution 9.23. The trigonometric form of a complex number tells us that

$$\begin{aligned}e^{i\theta} &= \cos\theta + i\sin\theta, \\ e^{i(-\theta)} &= \cos(-\theta) + i\sin(-\theta) \\ &= \cos\theta - i\sin\theta.\end{aligned}$$

From there, it is a matter of using the two equations to isolate $\cos\theta$ and $\sin\theta$ in this system of equations.

Solution 9.28. The complex number that we seek is

$$e^{i\frac{\pi}{2}} = \cos\frac{\pi}{2} + i\sin\frac{\pi}{2} = 0 + i\cdot 1 = i.$$

Then we can verify explicitly that

$$\arg(iz) \equiv \arg i + \arg z = \frac{\pi}{2} + \arg z.$$

Solution 9.29. Let $z = re^{i\theta}$. Then the proofs are straightforward using phase shift and reflection identities:

$$\begin{aligned}\arg(-z) &= \arg(-\cos\theta - i\sin\theta) = \arg(\cos(\pi+\theta) + i\sin(\pi+\theta)) \equiv \pi + \theta \equiv \pi + \arg z, \\ \arg(\overline{z}) &= \arg(\cos\theta - i\sin\theta) = \arg(\cos(-\theta) + i\sin(-\theta)) \equiv -\theta \equiv -\arg z.\end{aligned}$$

Solution 10.4. Let $f(x) = ax^2 + bx + c$ be a quadratic function.

1. Suppose $c = 0$. Then

$$f(x) = ax^2 + bx = x(ax+b).$$

This is equal to 0 if and only if one of x or $ax+b$ is equal to 0. Thus, the roots are 0 and $-\dfrac{b}{a}$.

2. Suppose $b = 0$. Then
$$f(x) = ax^2 + c.$$
Setting this equal to 0, we can isolate $x = \pm\sqrt{-\frac{c}{a}}$.

Solution 10.8. Let the coefficients a, b, c be rational. For one direction, suppose
$$\frac{-b + \sqrt{b^2 - 4ac}}{2a} = r_1,$$
$$\frac{-b - \sqrt{b^2 - 4ac}}{2a} = r_2$$
for rational numbers r_1 and r_2. Then
$$D = b^2 - 4ac = (a(r_1 - r_2))^2,$$
which is the square of a rational number. Conversely, suppose $D = b^2 - 4ac$ is the square of a rational number. Since we assumed that a, b, c are rational, $\frac{-b \pm \sqrt{b^2 - 4ac}}{2a}$ are also rational for either choice of the $\pm$ sign.

Solution 10.14. It is easy to verify that the identity holds by squaring both sides, which is a reversible step since both sides are non-negative. We leave this as an exercise to the reader. Instead, we will show how to find the decomposition in the first place.
Suppose there exist non-negative real numbers x and y such that
$$\sqrt{a \pm b\sqrt{c}} = \sqrt{x} \pm \sqrt{y},$$
where the two $\pm$ symbols represent the same sign. Squaring yields
$$a \pm b\sqrt{c} = (x + y) \pm 2\sqrt{xy}.$$
Inspired by how it is possible to equate real parts and equal imaginary parts when dealing with complex numbers, we make a non-rigorous guess that
$$a = x + y \quad \text{and} \quad b\sqrt{c} = 2\sqrt{xy}$$
will work. Since both sides are positive in the latter equation, the equation is equivalent to its squared counterpart $b^2c = 4xy$. So $x + y = a$ and $xy = \frac{b^2c}{4}$. By Vieta's formulas, x and y must be roots of the quadratic $f(z) = z^2 - az + \frac{b^2c}{4}$. By the quadratic formula, these roots are
$$x, y = \frac{a \pm \sqrt{a^2 - b^2c}}{2}.$$
As it turns out, these x and y indeed work out to form a decomposition $\sqrt{x} \pm \sqrt{y}$ that can be algebraically verified, as we mentioned at the beginning of the solution.

Solution 11.7. Non-constant linear polynomials have degree 1. According to Theorem 11.6,
$$\deg(f \circ g) = (\deg f) \cdot (\deg g)$$
if f and g are non-constant polynomials. The result follows, since multiplication by 1 leaves a number unchanged.

Solution 11.9. Let f be a polynomial with positive real coefficients. Suppose for contradiction that r is a non-negative real root of f. Then $f(r)$ computes to a positive real number, so r cannot be a root of f. Thus, if f has a real root, it must be negative.

Solution 11.10. Let

$$f(x) = a_n x^n + a_{n-1}x^{n-1} + \cdots + a_2x^2 + a_1x + a_0.$$

If $f(0) = 0$ then all but the constant term disappears and so $a_0 = 0$. Conversely, if $a_0 = 0$ then $f(0) = 0$ because all the non-constant terms have a factor of x.

Solution 11.18. Inspired by how we showed that the non-real complex roots of polynomials with real coefficients come in complex conjugate pairs in Theorem 11.16, we proceed as follows. Let f be a polynomial with rational coefficients, let c be a positive rational number that is not the square of a rational number, and let $r = a + b\sqrt{c}$ be a root of f, where a and b are rational numbers. Now let

$$r_1 = a_1 + b_1\sqrt{c},$$
$$r_2 = a_2 + b_2\sqrt{c},$$

where a_1, a_2 and b_1, b_2 are rational numbers. We leave it to the reader to show that

$$\overline{r_1 \cdot r_2} = \overline{r_1} \cdot \overline{r_2},$$
$$\overline{r_1 + r_2} = \overline{r_1} + \overline{r_2}.$$

Moreover, for any rational number t, we can show that $\overline{tr} = t\overline{r}$. Now the proof is exactly the same as the proof for showing that the non-real complex roots of polynomials with real coefficients come in complex conjugate pairs.

Solution 11.25. We perform long division of polynomials as follows:

$$\begin{array}{r|rrrrr}
 & & & \frac{3}{2}x^2 & -\frac{3}{4}x & -\frac{21}{8} \\ \hline
2x^2 + x + 4 & 3x^4 + & 0x^3 + & 0x^2 & + 10x & + 5 \\
 & -(3x^4 + & \frac{3}{2}x^3 + & 6x^2) & & \\ \hline
 & & -\frac{3}{2}x^3 - & 6x^2 & + 10x & \\
 & & -(-\frac{3}{2}x^3 - & \frac{3}{4}x^2 & - 3x) & \\ \hline
 & & & -\frac{21}{4}x^2 & + 13x & + 5 \\
 & & & -(-\frac{21}{4}x^2 & - \frac{21}{8}x & - \frac{21}{2}) \\ \hline
 & & & & \frac{125}{8}x & + \frac{31}{2}
\end{array}$$

Therefore the quotient is $\frac{3}{2}x^2 - \frac{3}{4}x - \frac{21}{8}$ and the remainder is $\frac{125}{8}x - \frac{31}{2}$.

Solution 11.31. Suppose, by contrapositive, that neither of p or q is the 0 polynomial. If pq has a root z, then $(pq)(z) = 0$. Then $p(z)q(z) = 0$ and z is a root of at least one of p or q. By Theorem 11.30, p has at most $\deg p$ distinct roots, and q has at most most $\deg q$ distinct roots. This means that there are at most $\deg p + \deg q$ values of z, which means pq takes on the value 0 for only finitely many complex numbers. By Theorem 11.19, pq is not 0 everywhere. By the contrapositive of what we just proved, if pq is 0 everywhere, then at least one of p or q is the zero polynomial.

Solution 11.42. Let $ax^3 + bx^2 + cx + d$ be a cubic equation, and let its roots be r_1, r_2, r_3. Then expanding the factored form of the equation yields

$$\begin{aligned} ax^3 + bx^2 + cx + d &= a(x - r_1)(x - r_2)(x - r_3) \\ &= ax^3 - a(r_1 + r_2 + r_3)x^2 + a(r_1r_2 + r_2r_3 + r_1r_3)x - ar_1r_2r_3. \end{aligned}$$

Comparing non-leading coefficients yields the formulas:

$$r_1 + r_2 + r_3 = -\frac{b}{a},$$
$$r_1r_2 + r_2r_3 + r_1r_3 = \frac{c}{a},$$
$$r_1r_2r_3 = -\frac{d}{a}.$$

We are allowed to match coefficients of polynomials that agree everywhere, thanks to Theorem 11.19.

Solution 11.52. We can rewrite the expression

$$r^2 + s^2 + t^2 = (r+s+t)^2 - 2(rs+st+rt) = \sigma_1^2 - 2\sigma_2.$$

But we know that $\sigma_k = (-1)^k \frac{a_{n-k}}{a_n}$, so $\sigma_1 = -\frac{a_2}{a_3}$ and $\sigma_2 = \frac{a_1}{a_3}$. The answer is

$$\sigma_1^2 - 2\sigma_2 = \frac{a_2^2}{a_3^2} - \frac{2a_1}{a_3} = \frac{a_2^2 - 2a_1a_3}{a_3^2} = \frac{3^2 - 2\cdot 2\cdot 5}{2^2} = -\frac{11}{4}.$$

Solution 11.56. Let $f(x,y,z) = (x+y+z)^3 - x^3 - y^3 - z^3$. Note that each of the scenarios $x = -y, y = -z, z = -x$ individually leads to $f(x,y,z) = 0$. This means each of the linear polynomials $x+y, y+z, z+x$ is a factor of f. So we can guess that there is a polynomial g such that

$$f(x,y,z) = (x+y)(y+z)(z+x)g(x,y,z).$$

Since $\deg f = 3$, we know that g must be a constant c. Now we can substitute constants into f, such as $(x,y,z) = (1,1,1)$ to get

$$c = \frac{(x+y+z)^3 - x^3 - y^3 - z^3}{(x+y)(y+z)(z+x)} = \frac{3^3 - 1^3 - 1^3 - 1^3}{2\cdot 2\cdot 2} = 3.$$

Indeed, expanding allows us to check that

$$(x+y+z)^3 - x^3 - y^3 - z^3 = 3(x+y)(y+z)(z+x).$$

Solution 11.58. We use the difference of squares, and difference and sum of cubes factorizations to get

$$x^6 - y^6 = (x^3)^2 - (y^3)^2 = (x^3 - y^3)(x^3 + y^3) = (x-y)(x^2 + xy + y^2)(x+y)(x^2 - xy + y^2),$$

where the quadratic factors are irreducible, as long as we are seeking real coefficients, because splitting them into linear factors using the quadratic formula yields complex coefficients.

Solution 11.59. Let ω be a non-trivial third root of unity. Then $\omega^3 = 1$, which tells us that

$$0 = \omega^3 - 1 = (\omega - 1)(\omega^2 + \omega + 1).$$

Since $\omega \neq 1$, we find that

$$\omega^2 + \omega + 1 = 0,$$

so a solution is $x^2 + x + 1$.

In general, this technique can be used to find a polynomial with degree less than n that is satisfied by an nth root of unity ω about which we have some additional information. It is always true that $\omega^n - 1$, but sometimes we are capable of producing a sharper polynomial (i.e. a factor of $\omega^n - 1$) and that might be necessary. For related material, see Volume 3.

Solution 11.60. Let $P(x) = \sum_{k=0}^{n} a_k x^k$. Then

$$P(a) - P(b) = \sum_{k=0}^{n} a_k(a^k - b^k).$$

By the difference of powers factorizations, $a - b$ divides $a^k - b^k$ for any positive integer k, and $a^0 - b^0 = 0$. Thus, $a - b$ divides $P(a) - P(b)$. In the case that b is a root of P, it means that $P(b) = 0$, so the result asserts that $a - b$ divides $P(a)$.

Solution 11.63. The idea is to replace each instance of c with $-(a + b)$ and to use the factorizations of $(a + b)^n - a^n - b^n$ for $n = 3, 5, 7$ (Theorem 11.62). For the first identity, this yields

$$\begin{aligned}\frac{f(5)}{f(3)} &= \frac{3}{5} \cdot \frac{a^5 + b^5 - (a + b)^5}{a^3 + b^3 - (a + b)^3} \\ &= \frac{3}{5} \cdot \frac{5ab(a + b)(a^2 + ab + b^2)}{3ab(a + b)} \\ &= a^2 + ab + b^2 \\ &= \frac{a^2 + b^2 + (a + b)^2}{2} = f(2).\end{aligned}$$

For the second identity,

$$\begin{aligned}\frac{f(7)}{f(5)} &= \frac{5}{7} \cdot \frac{a^7 + b^7 - (a + b)^7}{a^5 + b^5 - (a + b)^5} \\ &= \frac{5}{7} \cdot \frac{7ab(a + b)(a^2 + ab + b^2)^2}{5ab(a + b)(a^2 + ab + b^2)} \\ &= a^2 + ab + b^2 \\ &= \frac{a^2 + b^2 + (a + b)^2}{2} = f(2).\end{aligned}$$

Technically, we should not have divided by any quantities or cancelled any common factors between a numerator and denominator without being sure that the expressions do not assume a null value. However, the proof is cleaner this way and it is clear that a proof without division or cancellation could be produced in the same way by using the same factorizations.

Solution 12.1. By the trivial inequality,

$$a^2 + b^2 \geq 2ab \iff (a - b)^2 \geq 0.$$

Equality holds if and only of $a = b$. Adding the analogous inequalities $b^2 + c^2 \geq 2bc$ and $c^2 + a^2 \geq 2ca$ to it and dividing by 2 yields

$$a^2 + b^2 + c^2 \geq ab + bc + ca.$$

Equality holds if and only if $a = b = c$.

Solution 12.4. Using $S(n)$ and the stated substitution,

$$\begin{aligned}\frac{1}{n}\left(a_1 + a_2 + \cdots + a_{n-1} + \sqrt[n-1]{a_1 a_2 \cdots a_{n-1}}\right) &\geq \sqrt[n]{a_1 a_2 \cdots a_{n-1} \cdot \sqrt[n-1]{a_1 a_2 \cdots a_{n-1}}} \\ &= \left((a_1 a_2 \cdots a_{n-1})^{1+\frac{1}{n-1}}\right)^{\frac{1}{n}} \\ &= (a_1 a_2 \cdots a_{n-1})^{\frac{1}{n-1}} \\ a_1 + a_2 + \cdots + a_{n-1} + \sqrt[n-1]{a_1 a_2 \cdots a_{n-1}} &= n(a_1 a_2 \cdots a_{n-1})^{\frac{1}{n-1}} \\ a_1 + a_2 + \cdots + a_{n-1} &= (n - 1)(a_1 a_2 \cdots a_{n-1})^{\frac{1}{n-1}} \\ \frac{a_1 + a_2 + \cdots + a_{n-1}}{n - 1} &= \sqrt[n-1]{a_1 a_2 \cdots a_{n-1}}.\end{aligned}$$

As with the proof the theorem, the equality condition for $S(n-1)$ follows from that of $S(n)$.

Solution 12.6. The GM-HM inequality is equivalent to

$$\frac{\frac{1}{a_1}+\frac{1}{a_2}+\cdots+\frac{1}{a_n}}{n} \geq \sqrt[n]{\frac{1}{a_1}\cdot\frac{1}{a_2}\cdots\frac{1}{a_n}},$$

which follows from the AM-GM inequality. Using the equality condition for the AM-GM inequality, equality holds if and only if all the $\frac{1}{a_i}$ are equal, which is equivalent to all the a_i being equal. As with most proofs of multivariable inequalities, we have worked backwards here. This is fine, as long as the steps are reversible. Working backwards is an indispensable technique that only the most rigid pedant would avoid.

Solution 12.7. Both solutions will use the AM-GM inequality.

1. Let a and b be the dimensions of a rectangle with perimeter p, and let its area be A. By the AM-GM inequality for two variables,
$$A = ab \leq \left(\frac{a+b}{2}\right)^2 = \frac{p^2}{16}.$$
This upper bound for the area is reached if and only if $a = b = \frac{p}{4}$.

2. Let a, b, c be the dimensions of a box with surface area S, and let its volume be V. By the AM-GM inequality for three variables,
$$\frac{S}{6} = \frac{2ab+2bc+2ca}{6} = \frac{ab+bc+ca}{3} \geq (abc)^{\frac{2}{3}} = V^{\frac{2}{3}}.$$
So $V \leq \left(\frac{S}{6}\right)^{\frac{3}{2}}$. This upper bound for the volume is reached if and only if $ab = bc = ca = \frac{S}{6}$, which is true if and only if $a = b = c = \sqrt{\frac{S}{6}}$.

Solution 12.8. The trick is to use the factorization (see the list in Theorem 11.62)

$$\begin{aligned} x^3+y^3+z^3-3xyz &= (x+y+z)(x^2+y^2+z^2-xy-yz-zx) \\ &= \frac{1}{2}(x+y+z)((x-y)^2+(y-z)^2+(z-x)^2). \end{aligned}$$

By the trivial inequality, this is non-negative if and only if $x+y+z \geq 0$. Equality holds if and only if $x = y = z$ or $x+y+z = 0$.

Solution 12.11. By the Cauchy-Schwarz inequality,

$$\begin{aligned} (a_1^2+a_2^2+\cdots+a_n^2)^2 &= (a_1^2+a_2^2+\cdots+a_n^2)(a_{\sigma(1)}^2+a_{\sigma(2)}^2+\cdots+a_{\sigma(n)}^2) \\ &\geq (a_1a_{\sigma(1)}+a_2a_{\sigma(2)}+\cdots+a_na_{\sigma(n)})^2. \end{aligned}$$

By taking the square root of both sides, we get

$$a_1^2+a_2^2+\cdots+a_n^2 \geq |a_1a_{\sigma(1)}+a_2a_{\sigma(2)}+\cdots+a_na_{\sigma(n)}|.$$

Solution 12.12. A technique to keep in mind is that all of the numbers in one of the two n-tuples in Cauchy-Schwarz can be filled with (possibly unequal) constants. In this case, we can fill one of the n-tuples with all 1's to get

$$n(a_1^2+a_2^2+\cdots+a_n^2) = \underbrace{(1^2+1^2+\cdots+1^2)}_{n \text{ copies of } 1^2}(a_1^2+a_2^2+\cdots+a_n^2) \geq (a_1+a_2+\cdots+a_n)^2.$$

Since $1 \neq 0$, equality holds if and only if there exists a real number r such that $1 \cdot r = a_i$ for all i, which is equivalent to all the a_i being equal. This inequality leads to

$$\sqrt{\frac{a_1^2 + a_2^2 + \cdots + a_n^2}{n}} \geq \left| \frac{a_1 + a_2 + \cdots + a_n}{n} \right| \geq \frac{a_1 + a_2 + \cdots + a_n}{n}.$$

Equality holds in the left inequality if and only if all of the a_i are equal and equality holds in the right inequality if and only if $a_1 + a_2 + \cdots + a_n$ is non-negative. Thus, equality holds in the RMS-AM inequality if and only if all of the a_i are non-negative and equal.

Solution 12.13. By Engel's form of the Cauchy-Schwarz inequality (Corollary 12.10),

$$\begin{aligned}\frac{a}{b+c} + \frac{b}{c+a} + \frac{c}{a+b} &= \frac{a^2}{ab+ca} + \frac{b^2}{bc+ab} + \frac{c^2}{ca+bc} \\ &\geq \frac{(a+b+c)^2}{(ab+ca)+(bc+ab)+(ca+bc)} \\ &= \frac{a^2+b^2+c^2}{2(ab+bc+ca)} + 1.\end{aligned}$$

By Problem 12.1, we know that

$$a^2 + b^2 + c^2 \geq ab + bc + ca,$$

so it holds that

$$\frac{a^2+b^2+c^2}{2(ab+bc+ca)} + 1 \geq \frac{3}{2}.$$

We know that equality holds in the last step if and only if $a = b = c$, so this is a necessary criterion for equality in Nesbitt's inequality. Substituting $a = b = c$ into Nesbitt's inequality shows that this condition is also sufficient.

Solution 12.16. By substituting the stated formulas into Euler's inequality and clearing the denominators, we find that we want to prove that

$$abc \geq (-a+b+c)(a-b+c)(a+b-c).$$

This is another way of writing Schur's inequality for $t = 1$. We leave it to the reader to watch the solution unfurl by expanding the right side of this equality and by expanding Schur for $t = 1$. Since a, b, c are side lengths of a triangle, none of them can be 0, so the only case of equality is $a = b = c$.

Solution 12.22. We will use Jensen's inequality. For the interval $(0, \infty)$ on which the following functions in the variable x are defined, we observe the strict convexity of x^t with a fixed real $t > 1$, the strict concavity of x^t with $0 < t < 1$, and the strict convexity of x^t for $t < 0$. Now we break the argument into three cases, according to where r and s lie relative to 0:

1. Suppose $r > s > 0$. Then $\frac{r}{s} > 1$, so $f(x) = x^{\frac{r}{s}}$ is strictly convex. By Jensen,

$$(\lambda_1 x_1^s + \lambda_2 x_2^s + \cdots + \lambda_n x_n^s)^{\frac{r}{s}} \leq \lambda_1 x_1^r + \lambda_2 x_2^r + \cdots + \lambda_n x_n^r.$$

   Taking both sides to the power of $\frac{1}{r}$ preserves the direction of the inequality because $g(x) = x^{\frac{1}{r}}$ is an increasing function for $r > 0$.

2. Suppose $r > 0 > s$. Then $\frac{r}{s} < 0$, so $f(x) = x^{\frac{r}{s}}$ is strictly convex. The same inequality holds again, and taking both sides to the power of $\frac{1}{r}$ still preserves the inequality because $g(x) = x^{\frac{1}{r}}$ is increasing again, due to $r > 0$.

3. Suppose $0 > r > s$. Then $0 < \frac{r}{s} < 1$, so $x^{\frac{r}{s}}$ is strictly concave this time. Then the reverse of the inequality from the previous two cases holds, but this is reversed again by taking both sides to the power of $\frac{1}{r}$ because $g(x) = x^{\frac{1}{r}}$ is decreasing for $r < 0$.

By Jensen, equality holds if and only if

$$x_1^s = x_2^s = \cdots = x_n^s,$$

since all of the λ_i are positive; in particular, none of the λ_i are 0. Since $s \neq 0$, we can invert it to state that equality holds and only if all of the x_i are equal.

Solution 12.24. We will use weighted AM-GM. To this end, let $\lambda_i = \frac{1}{\mu_i}$ for each index $i \in [n]$ so that $\lambda_i \in (0, 1)$, and

$$\lambda_1 + \lambda_2 + \cdots + \lambda_n = 1$$

By the weighted AM-GM inequality,

$$x_1 x_2 \cdots x_n = (x_1^{\mu_1})^{\lambda_1}(x_2^{\mu_2})^{\lambda_2} \cdots (x_n^{\mu_n})^{\lambda_n} \leq \lambda_1 x_1^{\mu_1} + \lambda_2 x_2^{\mu_2} + \cdots + \lambda_n x_n^{\mu_n} = \frac{x_1^{\mu_1}}{\mu_1} + \frac{x_2^{\mu_2}}{\mu_2} + \cdots + \frac{x_n^{\mu_n}}{\mu_n}.$$

None of the $\frac{1}{\mu_i}$ can be 0, so the equality criterion of weighted AM-GM says that equality holds if and only if $x_1^{\mu_1} = x_2^{\mu_2} = \cdots = x_n^{\mu_n}$.

Solution 12.26. In Hölder's inequality, we take $m = 2$ and the two lists to be $(x_k^p)_{k=1}^n$ and $(y_k^q)_{k=1}^n$. From there, the result easily follows because, for any real t, $(t^p)^{\frac{1}{p}} = t$ and $(t^q)^{\frac{1}{q}} = t$. By the equality criterion of Hölder's inequality, equality holds if and only if one of the lists, that is $(x_k^p)_{k=1}^n$ or $(y_k^q)_{k=1}^n$, consists of all 0's or

$$x_1^p : x_2^p : \cdots : x_k^p = y_1^q : y_2^q : \cdots : y_k^q.$$

So each list is a scaled version of the other, which establishes the existence of a scale factor c, which must be positive because all of the x_k and y_k are positive. To be precise about what we mean by a scale factor, it means $y_k^q = cx_k^p$ for all indices $k \in [n]$.

Solution 12.31. Fixing any positive integer n and taking the assumption for granted, we will prove the conclusion by induction on $k \in [n-1]$. In the case base, where $k = 1$, we prove the following by taking reversible steps:

$$\frac{a_2 - a_0}{2} \geq \frac{a_1 - a_0}{1}$$
$$a_2 - a_0 \geq 2a_1 - 2a_0$$
$$\frac{a_2 + a_0}{2} \geq a_1,$$

which is true by assumption. Suppose the conclusion is true for some integer k such that $1 \leq k < n - 1$. To get to $k + 1$, we first manipulate the induction hypothesis to get

$$\frac{a_{k+1} - a_0}{k+1} \geq \frac{a_k - a_0}{k}$$
$$k(a_{k+1} - a_0) \geq (k+1)(a_k - a_0)$$
$$ka_{k+1} + a_0 \geq (k+1)a_k.$$

Then, by assumption and the above,

$$\begin{aligned}
2a_{k+1} &\leq a_k + a_{k+2} \\
&\leq \frac{k}{k+1} \cdot a_{k+1} + \frac{1}{k+1} \cdot a_0 + a_{k+2} \\
2(k+1)a_{k+1} &\leq ka_{k+1} + a_0 + (k+1)a_{k+2} \\
(k+2)a_{k+1} &\leq (k+1)a_{k+2} + a_0 \\
(k+2)a_{k+1} - (k+2)a_0 &\leq (k+1)a_{k+2} - (k+1)a_0 \\
\frac{a_{k+1} - a_0}{k+1} &\leq \frac{a_{k+2} - a_0}{k+2},
\end{aligned}$$

as desired, which can be discovered by working backwards.

List of Symbols

Arithmetic

$\mathbb{Z}$	integers
$\mathbb{Z}_+$	positive integers
$\mathbb{Z}_{\geq 0}$	non-negative integers
$\mathbb{Q}$	rational numbers
$\mathbb{Q}_+$	positive rationals
$\mathbb{Q}_{\geq 0}$	non-negative rationals
$\mathbb{R}$	real numbers
$\mathbb{R}_+$	positive reals
$\mathbb{R}_{\geq 0}$	non-negative reals
$\mathbb{C}$	complex numbers
$\mathbb{F}$	field
$\pm$	plus or minus
$<, >$	strict inequality
$\leq, \geq$	non-strict inequality

Constants

$\zeta_k = e^{\frac{2k\pi}{m}i}$	m^{th} root of unity
π	pi
e	Euler's constant
ϕ	the golden ratio
i	the square root of -1

Functions

$\text{Dom}(f)$	domain
$\text{Rng}(f)$	range
$\lfloor \cdot \rfloor$	floor function
$\lceil \cdot \rceil$	ceiling function
sgn	signum function
max	maximum function
min	minimum function
det	determinant
Id_S	identity function on S
$f \circ g$	function composition
$n!$	factorial
$\overline{z}$	complex conjugate
$\underline{r}$	radical conjugate
σ	permutation
csc	cosecant
sin	sine
cos	cosine
sec	secant
cot	cotangent
tan	tangent
$\bullet$	dot product of vectors
$\times$	cross product of vectors

Logic

$\neg$	negation
$\vee$	disjunction, or
$\wedge$	conjunction, and
$\oplus$	exclusive or, XOR
$\implies$	implication
$\iff$	biconditional
$\equiv$	logical equivalence

Miscellaneous

$\exists$	existential quantifier
$\forall$	universal quantifier
$(a_i)_{i \in I}$	sequence indexed by I
$\sum$	summation notation
$\prod$	product notation
$a \sim b$	equivalence relation

Sets

$\varnothing$	empty set

$\in$	element of
$\notin$	not element of
$[n]$	$\{1, 2, \ldots, n\}$ for $n \in \mathbb{Z}_+$
$[n]^*$	$\{0, 1, 2, \ldots, n\}$ for $n \in \mathbb{Z}_{\geq 0}$
S^c	set complement
$\cup$	set union
$\cap$	set intersection
$A \backslash B$	set difference
$A \times B$	Cartesian product of sets
A^n	$\underbrace{A \times A \times \cdots \times A}_{n \text{ copies of } A}$
$\mathcal{P}(A)$	power set
$A \oplus B$	set symmetric difference
$\subseteq$	subset
$\subsetneq$	proper subset
$\supseteq$	superset
$\mathcal{U}$	universal set

Bibliography

"When we can't think for ourselves, we can always quote."

– *Ludwig Wittgenstein*

"Have you ever observed that we pay much more attention to a wise passage when it is quoted than when we read it in the original author?"

– *Philip Gilbert Hamerton*

"Everything of importance has been said before by somebody who did not discover it."

– *Alfred North Whitehead*

[1] David A. Cox at al. *Ideals, Varieties, and Algorithms, fourth edition.* Springer International Publishing, 2015, pp. 347–348.

[2] Radmila Manfrino et al. *Inequalities: A Mathematical Olympiad Approach.* Birkhauser Verlag AG, 2009, pp. 43–47.

[3] Titu Andreescu and Zuming Feng. *A Path to Combinatorics for Undergraduates: Counting Strategies.* Birkhauser, 2004, p. 164.

[4] Edward Barbeau and Samer Seraj. "Sum of cubes is square of sum". In: *Notes on Number Theory and Discrete Mathematics* 19(1) (2013), pp. 1–13.

[5] Evan Chen. *Euclidean Geometry in Mathematical Olympiads.* MAA Press, 2016, pp. 95–118.

[6] Ronald Graham, Donald E. Knuth, and Oren Patashnik. *Concrete Mathematics, second edition.* Addison-Wesley, 1994, pp. 71–72.

[7] Pham Kim Hung. *Secrets in Inequalities: Advanced Inequalities.* GIL Publishing House, 2008.

[8] Pham Kim Hung. *Secrets in Inequalities: Basic Inequalities.* GIL Publishing House, 2007.

[9] Samer Seraj. "A Short Proof of Euler's Inequality". In: *Resonance - Journal of Science Education* 20(1) (2015), p. 75.

[10] J. Michael Steele. *The Cauchy-Schwarz Masterclass: An Introduction to the Art of Mathematical Inequalities.* Cambridge University Press, 2004.

Index

"We raise to degrees (of wisdom) whom We please: but over all endued with knowledge is one, the All-Knowing."

– *Qur'an 12:76*

About the Author

"If you would be a real seeker after truth, it is necessary that at least once in your life you doubt, as far as possible, all things."

– *René Descartes, Principia Philosophiae*

"He who has a why to live for can bear almost any how."

– *Friedrich Nietzsche, Twilight of the Idols*

Samer Seraj is the owner of Existsforall Academy, where he enjoys teaching mathematics to his students. He earned his undergraduate degree in mathematics from the University of Toronto with membership at Trinity College. He also obtained a Masters degree in international affairs from King's College London. He is the author of several mathematical textbooks and has published research papers in journals. His vacation time is often spent on marking math contests and giving invited lectures, including formerly as a trainer of Math Team Canada for the International Mathematical Olympiad.

https://existsforall.com/

Made in the USA
Monee, IL
05 November 2026

67378949R00109